DOWN TO THE WIRE:
STUDIES IN THE DIFFUSION AND REGULATION OF TELECOMMUNICATIONS TECHNOLOGIES

DOWN TO THE WIRE:
STUDIES IN THE DIFFUSION AND REGULATION OF TELECOMMUNICATIONS TECHNOLOGIES

ALLAN L. SHAMPINE (EDITOR)

Nova Science Publishers, Inc.
New York

Senior Editors: Susan Boriotti and Donna Dennis
Coordinating Editor: Tatiana Shohov
Office Manager: Annette Hellinger
Graphics: Wanda Serrano
Editorial Production: Marius Andronie, Maya Columbus, Vladimir Klestov,
Matthew Kozlowski and Tom Moceri
Circulation: Ave Maria Gonzalez, Vera Popovich, Raymond Davis, Melissa Diaz,
Magdalena Nuñez, Marlene Nuñez and Jeannie Pappas
Communications and Acquisitions: Serge P. Shohov
Marketing: Cathy DeGregory

Library of Congress Cataloging-in-Publication Data

Down to the wire: studies in the diffusion and regulation of telecommunications technologies /
Allan L. Shampine, editor.
p. cm.
Includes index.
ISBN: 1-59033-778-6 (hardcover)
1. Telecommunication. 2. Telecommunication—Law and legislation.
3. Telecommunication policy. I. Shampine, Allan L.

TK5101.D66 2003
384—dc21 2003012604

400 Oser Ave, Suite 1600
Hauppauge, New York 11788-3619
Tele. 631-231-7269 Fax 631-231-8175
e-mail: Novascience@earthlink.net
Web Site: http://www.novapublishers.com

Printed in the United States of America

CONTENTS

OVERVIEW[*]

Allan L. Shampine
Lexecon, Inc.

In recent years, regulators and businesses have struggled to keep pace with an onslaught of new technologies that has dramatically changed the world's telecommunications networks. This volume gathers together research intended for those who must grapple with these changes – the people who must decide whether and how to deploy new telecommunications technologies, and those who must regulate the technologies. The research is divided into three sections focusing on the past, present and future of telecommunications innovations. The first section, Lessons from the Past, examines what lessons concerning adoption and regulation can be learned by studying the diffusion of recent telecommunications technologies. The second section, Decision Making with Respect to Diffusing Technologies, examines possible improvements to the decision-making processes used by companies and regulators in the face of limited information and dynamic markets. The final section, Public Policy and Future Diffusion, examines public policy implications for future technologies.

LESSONS FROM THE PAST

As the wave of recent bankruptcies in the telecommunications industry has made clear, predicting the success of new technologies is difficult. Firms with similar levels of investment and availability of service often have very different experiences with respect to adoption. Examples of technologies that were both deployed rapidly in the United States but had different adoption rates are digital wireless and high-speed Internet access services. On the one hand, digital wireless services were quickly adopted. In 1997 the great majority of wireless customers used analog services, and digital networks were only beginning to be

---

[*] Any edited volume depends on talented contributors, and this one is no exception, so my thanks go first and foremost to the authors of the papers contained herein. I would also like to thank my colleague Hal Sider, who first introduced me to the economics of telecommunications and with whom I have worked on these issues for many years.
The opinions contained in this volume are those of the individual authors and do not necessarily reflect those of their employers or of their fellow authors.

deployed using newly allocated spectrum or network overlays. As of June 1997, there were 46 million wireless subscribers, of whom 93 percent used analog services. As of June 2002, there were over 126 million wireless subscribers, 87 percent of whom used digital services. Only 16 million analog customers remained.[1] The stocks of wireless companies have done relatively well in the recent market downturn, in part because of the success of digital wireless service.

On the other hand, high-speed Internet access technologies have had relatively slow adoption rates in the United States. This slow adoption appears to have little to do with availability. Companies have had wide coverage and still had such low adoption rates as to end up in bankruptcy. Teligent, for example, launched its service in October 1998 and claimed to offer wireless high-speed data services to over 100 million people within fourteen months.[2] Despite the accumulation of over $1 billion in network assets, they filed for bankruptcy in mid-2001.[3] Similarly, Winstar spent billions of dollars on a wireless data network and ended up in bankruptcy. Its assets, with a book value of $5 billion, were auctioned for $42.5 million in 2001.[4] Companies offering high-speed Internet access via DSL have also fared poorly. DSL services are widely available, with 54 percent of households in the U.S. having access to DSL in 2002,[5] but overall DSL penetration remains low. Indeed, the three largest independent DSL operators (NorthPoint, Rhythms, and Covad) have all ended up in bankruptcy. NorthPoint's assets, with a book value of $479 million, were sold to AT&T for $135 million.[6, 7] WorldCom acquired key parts of Rhythm's assets, with a book value of $172 million, for $31 million.[8] The third company, Covad Communications, has emerged from bankruptcy. However, despite offering DSL service to more than 40 million homes and businesses, Covad still has only 381,000 active lines – a penetration rate of less than 1 percent.[9]

This sort of misestimation has had serious consequences for the economy. Data from Bankruptcy DataSource illustrate the magnitude of the problem.[10] In 1998, 3 telecommunications companies with $4.1 billion in assets filed for bankruptcy. In 1999 and 2000, 8 and 7 companies respectively filed with roughly $5.5 billion in assets each year. In 2001, 21 telecommunications companies with $18.7 billion in assets filed for bankruptcy, and in 2002, 20 companies with an astonishing $176 billion in assets filed. The 2002 figure includes the filing by WorldCom with $104 billion in assets. Excluding WorldCom, the other 19 bankrupt companies had $72 billion in assets, or roughly 13 times the levels of 1999 and 2000.

---

[1] Cellular Telecommunications & Internet Association (CTIA), "CTIA's Wireless Industry Indices Semi-Annual Data Survey Results," December 2002, p.48.

[2] Teligent Corporate Profile, 2000, p.1.

[3] Bankruptcy DataSource.

[4] Winstar press release, "IDT Corp. Announces the Acquisition of Winstar Communications, Inc.," December 20, 2001.

[5] The Yankee Group, Wireless Voice and Broadband Access unlock Local Line's Grip on Consumers, 10/02, p.5.

[6] AT&T Press Release, "AT&T Completes Acquisition of Assets of NorthPoint Communications," May 25, 2001.

[7] AT&T has subsequently written down the value of its DSL assets by $200 million. AT&T Press Release, "AT&T to Take Restructuring and Asset Impairment Charges," January 6, 2003.

[8] WorldCom Press Release, "WorldCom Closes Rhythms Transaction," December 5, 2001.

[9] Covad press release, "Covad announced fourth quarter operating statistics," January 14, 2003.

[10] Bankruptcy DataSource reports on bankruptcy filings by relatively large companies, i.e., those with more than $50 million in assets as of their bankruptcy filing, so it shows only relatively high profile bankruptcies.

Researchers have underestimated demand just as often, although the costs of doing so tend to be lower. Crandall and Jackson in this volume include a survey of new technologies whose adoption rates were underestimated. For example, most forecasts of cellular adoption grossly underestimated the rate of adoption. Furthermore, it is not immediately obvious why the same technologies experience substantially different availability (infrastructure diffusion) and acceptance (consumer adoption) rates in different countries, and even in different areas within the same country. United States regulators with dashed hopes of universal high-speed Internet access have been particularly interested in the latter question as they ask how South Korea can have a DSL penetration rate nearing 75 percent[11] while the United States' rate is closer to 12 percent.[12]

The papers in the first section of this volume (Lessons from the Past) examine disparities in diffusion within and between countries in the hope of providing some insights into how these disparities might be remedied should policy makers desire to do so. In the first paper of this section, Greenstein, Lizardo and Spiller examine regional disparities in U.S. adoption of advanced information infrastructure. They evaluate telecommunications and computing capital stocks by developing data series on fiber optic cable deployment, the number of large-scale computer (or "server") users, and the processing capacity of servers in a given area for the late 1980s and early 1990s. They find persistent regional disparities favoring dense urban areas, rich areas and regions with large amounts of white collar work. They also find that the relative differences have decreased over time. They conclude that the data do not support the argument that the diffusion of advanced IT exacerbated regional disparities in economic development in the United States. They further suggest that decreases in regional disparities may be cause for regulators to revisit universal service provisions for related technologies.

While the deployment of infrastructure documented by Greenstein, Lizardo and Spiller was sufficient to enable nearly ubiquitous dial-up Internet access within the United States, higher infrastructure requirements of new technologies may increase regional service disparities once again. For example, certain graphical and data services are only usable with relatively high data transfer rates – rates that may or may not be achievable with the existing infrastructure in some regions, such as rural areas. Other countries have also had differing experiences with infrastructure disparities.

The next paper, by Shampine, examines international disparities in the adoption of central office digital switching and contrasts the results with cross-company disparities within the United States. The author concludes that U.S. firms with large networks were slower to adopt digital switching (primarily because of their large installed bases of analog equipment) and were therefore 'leap-frogged' by other companies, but that this 'leap-frog' effect does not appear in international comparisons. Countries which were slow in deploying analog infrastructure also tended to be slow in deploying digital infrastructure. Indeed, the existing levels of infrastructure and GDP are among the strongest national determinants of the rate of digital adoption. This contrast between inter-company and inter-country results suggests that the "digital divide" between countries may be more persistent than disparities within a given country.

---

[11] Alan Hellawell, "Korea Portfolio Strategy," Lehman Brothers, February 19, 2003.

[12] The U.S. rate is for all high-speed Internet access, based on 12.8 million high-speed lines in the U.S. as of December 2002 and 105 million households. FCC, High-Speed Services for Internet Access: Status as of December 31, 2001 (July 2002). Bureau of the Census, Current Population Reports, America's Families and Living Arrangements, June 2001.

The final paper in the first section, by Wayne Fu, examines regulatory distortions as an influence on wireless company adoption and pricing in the Taiwan mobile phone industry. Wireless service in Taiwan was only available from the monopolist incumbent until 1998, when eight entrants were awarded spectrum. The regulatory authority mandated interconnection at TELRIC (total element long run incremental cost) rates as pioneered by the United States' FCC. However, the author finds that the regulators and firms did not properly execute TELRIC, resulting in interconnection fees set well in excess of cost. Since the industry uses a caller-pays system that passed along these fees, consumers have tended to join the largest networks (a band-wagon effect) to minimize their costs, resulting in increasing concentration over time and regulatory concerns about "tipping," where a few networks come to dominate the market.

These papers suggest that the determinants of adoption vary between countries and technologies, and that the level of regional variation may depend on a variety of factors. These variations may make it difficult to predict demand based on the experiences of other technologies. This conclusion is supported by the business failures discussed previously and an examination of historical forecasts.[13] There does not appear to be a "one size fits all" method for forecasting or regulating telecommunications innovations. While academics can always wait for more data, companies and regulators do not have that luxury. How, then, should they make decisions given such uncertainty? The next section in this book attempts to shed some light on this question.

Decision Making with Respect to Diffusing Technologies

The second section in this volume addresses how companies and regulators can make better decisions with limited information. The usual problem with a new technology is that there is little basis for predicting demand. Even technologies that simply improve upon an already existing technology (such as high-speed Internet access) may be priced in a previously unexplored area of the demand curve. The failure of many DSL providers demonstrates how costly overestimating the demand for a product can be.

Companies expend substantial resources attempting to estimate demand elasticities and project future demand. In the cases where market transactions are unavailable, companies often turn to contingent valuation – that is, they ask people to value things or choose between hypothetical bundles. There are many concerns with this sort of study, but the methods persist simply because, in the absence of market data, most businesses prefer some number to no number. Generally, the trick is to extract a reasonable demand curve from the survey data. One common problem with doing so is the fact that people tend to respond to questions using certain informational shorthands, such as rounding their willingness to pay to the nearest multiple of ten ($30, $40, etc.). Rappoport, Taylor and Kridel offer a new solution to this problem. They use survey (contingent-valuation) data on DSL and cable modem willingness-to-pay in order to estimate elasticities, but they do so by estimating a kernel density function as an approximation to the empirical frequency distribution of the willingness-to-pays and

[13] See Crandall and Jackson in this volume for a discussion of how historical forecasts underestimated cellular adoption.

then calculate elasticities at various points on the function. This method allows a continuous approximation to the frequency distribution that avoids problems with "pile-up" points and other data concerns.[14] Application of the method provides estimates of high-speed Internet access elasticities similar to those found in previous studies using other methods. The robustness of the results suggests that the method may be a useful addition to a researcher's toolbox.

The next article provides an example of how a regulated monopolist evaluated and responded to a new competing technology. In 2001, the United States Postal Service observed a decline in mail volume that it attributed to diversion of mail to the Internet. The Postal Service was concerned about the impact of the Internet on its offerings and commissioned RCF Consulting to study the matter. Bernstein presents the results of that analysis as a case study of how a regulated monopolist evaluated, forecasted and responded to emerging competition from a new technology. The results were intended to guide the Postal Service's decision-making process, but the methods are instructive for other companies or regulators who are interested in a detailed description of how one company actually went about evaluating and responding to a new technology.

Regulators are constantly faced with similar situations as they evaluate new technologies that are not addressed by existing legislation. This is not a simple process. Regulations evolve over time along with the industry, but not always at the same speed. During the process of updating regulations to be consistent with technological changes, new technologies and services may fall under different regulatory regimes due to historical factors. As a result, functionally equivalent services may be regulated differently because of historical distinctions between service providers or infrastructure. For example, cable and DSL high-speed Internet access services in the United States are regulated differently because cable companies operate under a different regulatory regime than local exchange carriers. Nonetheless, cable and DSL services are interrelated, and a coherent public policy approach to broadband diffusion in the United States requires unified regulation. The final paper in this section provides an example of the importance of such a unified approach.

In this paper, Burnstein and Aron examine the relationship between DSL and cable adoption. They assemble data on the availability of cable and DSL high-speed Internet access in the United States and compare them with geographic data on adoption levels. They find that the presence of intermodal competition (that is, the availability of both cable and DSL service) has a substantial and statistically significant positive effect on adoption. They conclude that a policy which promotes universal access by encouraging investment in one technology over another may therefore be less efficient than efforts to promote intermodal competition.

PUBLIC POLICY AND FUTURE DIFFUSION

The final section in this volume looks to the future. Given the lessons of the past, what can policy makers do to accelerate adoption of future technologies? Not surprisingly, there are few if any simple answers to such a question. A particular technology may be new, but it

[14] Responses are typically clustered at "pile-up" points such as even dollar amounts, or round numbers such as $10, $20, and so forth.

may well use portions of a legacy monopolist network, or it might require modifications of the existing network that could impact interconnection or resale regulations. The basic problem is that new telecommunications technologies and services do not operate in isolation. Increasingly, technologies blur together. For example, circuit switched networks are being replaced with packet switched networks that treat all traffic as data regardless of the source or application. The papers in this section discuss how regulators can best further public policy by encouraging the deployment and adoption of new technologies while still fulfilling their duties in regulating legacy services.

The first paper in this section illustrates the maxim that everything old is new again. Economides crafted his critique of the efficient component pricing rule (ECPR) during the initial discussion of unbundled network elements (UNEs) in the late 1990s. The FCC eventually ruled that UNEs should be priced based on total element long run incremental cost, and left the actual pricing to the states. The matter has been in litigation ever since with one trip to the Supreme Court so far. The paper is included here both as an important historical commentary and as a discussion of the current issues being debated before the FCC. While the ECPR is no longer an issue, the underlying concerns of private versus public opportunity costs and their roles in pricing are still being debated, and Economides' critique remains highly relevant.

As just noted, there has been enormous debate about the proper way to measure forward-looking economic costs (TELRIC) for existing infrastructure. Evaluations of new technologies can be even more difficult since there is little if any existing infrastructure to study and their full scope may not be known. While this uncertainty must be taken into account, the benefits of new technologies can be substantial and should not be ignored simply because they are hard to predict. In the next paper of this section, Crandall and Jackson demonstrate how substantial the benefits can be as they estimate the economic benefits of accelerated adoption of high-speed Internet access. They do so using two different methods. First, they estimate a prospective demand function and calculate the expected consumer surplus associated with different levels of adoption. Second, they attempt to calculate the consumer surplus of some of the benefits of high-speed Internet access, including retail, transportation, home entertainment and health care. The results of the two methods are similar. Crandall and Jackson estimate that the net present value of a faster rollout of high-speed access to be around $500 billion, with annual benefits reaching $300 billion. The magnitude of this effect emphasizes the importance of appropriate regulation.

In the final volume of this paper, Bauer, Kim and Wildman examine how regulators can take into account regional and technological variations when setting public policy with respect to high-speed Internet access. They examine how the spatial structure of costs and benefits from broadband networks will impact the ability of private industry to achieve a socially optimal outcome, and what level of regulatory intervention (local, inter-local or global) would be most appropriate for a given combination. Their results are consistent with the disparities observed in the other papers in this volume. That is, even for a given technology, the costs and benefits of adoption may vary by region, leading to different levels of deployment and adoption, and thus requiring different types and levels of regulation.

CONCLUSION

Investors, regulators and academics have had difficulty with new technologies as recent bankruptcies in the telecommunications industry and regulatory squabbles demonstrate, but there is reason for optimism. The importance of telecommunications to economic development is widely recognized. As a result, both regulators and industry sources have become increasingly aggressive in gathering data on new technologies at earlier stages. Much of the work in this volume stems from these data gathering efforts. Furthermore, this research, and ongoing research like it, demonstrates that the problems, while difficult, are tractable. Reasonable projections can be made, and reasonable conclusions can be drawn given currently available data. Of course, new technologies are just that – new. Truly groundbreaking advances are, by their very nature, different from their predecessors.

Luckily, regulators are aware of the speed with which new telecommunications technologies diffuse, and are making efforts to see that regulations provide the proper incentives for innovation and new investment. Companies have also learned some painful lessons from the recent recession and the resulting volatility of infrastructure values. Hopefully the research contained within this volume will help both groups in their efforts as the industry continues to evolve. If history is any indication, it will be an exciting trip.

PART I.
LESSONS FROM THE PAST

Chapter 1

THE EVOLUTION OF ADVANCED LARGE SCALE INFORMATION INFRASTRUCTURE IN THE UNITED STATES*

Shane M. Greenstein
Kellogg Graduate School of Management
Northwestern University
Mercedes M. Lizardo
Instituto Tecnologico de Santo Domingo
Dominican Republic
Pablo T. Spiller
Haas School of Business
University of California, Berkeley

I. INTRODUCTION

While studies have examined the regional development of information technology producers in the U.S.,[1] comparatively few have examined the regional diffusion of information technology to users. While a number of studies examine the returns from the use of information technology within organizations[2], little is known about the geographic concentration of those returns. This is a large gap in understanding how innovation translates into economic prosperity. Economic growth and advances in welfare arises from the use and adoption of technology in economic processes. Disparities between enterprises in the levels

* We would like to thank the Institute for Government and Public Affairs at the University of Illinois and the National Science Foundation for funding. We are grateful for comments received from Tim Bresnahan, Allan Shampine Steve Lanning, Mike Mazzeo, Greg Rosston, Walter Sosa, anonymous reviewers, and participants at conferences on Convergence in Telecommunications at UC Berkeley, Global Network 97 in Calgary, and 1997 Telecommunications Policy Research Conference in Alexandria, VA.

[1] For example, see Markusen et al [1986], Saxenian [1994], Steinmueller [1996], Forman, Goldfarb and Greenstein, [2002].

[2] For example, see Bresnahan and Greenstein [1997], and Brynolfsson and Hitt [1997].

and rate of adoption of new technology may lead to related disparities in regional economies, especially when different types of spillovers are limited by the proximity of economic activity.

In this study we empirically characterize the distribution of use of advanced information technology across the US. Here "advanced" means computing and communication technology near the technical frontier, whose increasing use in the 1980s enabled new goods and services and transformed downstream economic activity. We measure and analyze changes to the geographic distribution of advanced information technology (IT) in the late 1980s and early 1990s, the period just prior to the commercialization of the Internet. There was an explosion of commercial investment in information infrastructure in the late 1990s, but little is know about investment patterns prior to that.[3]

We first characterize the pattern of disparities between regions in the levels of advanced information infrastructure and access to it. We expect that economies of density in the provision and employment of many communication technologies give users strong incentives to locate these technologies in urban areas. Also, some industries are more intensive users of advanced information technology. Administrative-intensive work, such as banking, finance, insurance, real estate, or business services, tends to make heavy use of advanced computing and digital communications. Information infrastructure will tend to be found in the same places as these industries. Thus, we investigate how much of the location of information infrastructure can simply be explained by the density of human settlement and the location of industry.

We then investigate whether the diffusion of advanced IT follows an increasing "agglomeration" pattern or an increasing "dispersion" pattern over time. In an increasing agglomeration pattern, one or several geographically localized economic forces leads to increasing concentration of advanced infrastructure in a few places. In this situation the geographic areas that were early leaders continue to retain over time their relative lead over other regions and even build on it. In the increasing dispersion pattern, the opposite occurs. The geographic areas that were early leaders in technology lose their relative lead or cease to be leaders -- perhaps as part of a general re-ordering of the relative standing of regions.

Questions about the dispersion and agglomeration of investment are particularly interesting in light of the widespread investment in the late 1990s. In less than a decade after the commercialization of the Internet the supply and use of Internet infrastructure became nearly geographically ubiquitous.[4] Was this remarkable period of growth in infrastructure predeeded by dispersed investments in an earlier era?

We employ methods from the literature on economics of infrastructure in developing countries, suitably adapted to examine the regional distribution of advanced information infrastructure in a developed country. Our data on information technology capital at public and private enterprises come from surveys of users' information capital, which we aggregate to the smallest possible geographic unit. From these sub-indices we compute summaries of advanced IT and access to it. Following the spirit of the infrastructure literature, we interpret our indices as broad indicators of activities associated with building public and private digital IT networks in a region – i.e., development of labor markets for IT services, related

[3] For more on geography and growth of the Internet, see Cairncross [1997], Kolko [2002], Greenstein [2000], Kotkin [2000], Moss and Townsend [1997], Zooks [2000], and Forman, Goldfab and Greenstein [2002].

[4] See e.g., the analyses in Greenstein [2000], Downes and Greenstein [2002], Forman 2002], and Forman, Goldfarb and Greenstein [2002].

downstream applications, and localized inventive activity closely linked to advanced infrastructure.

We analyze three key ingredients of digital infrastructure: a) fiber optic deployment by local telephone companies, b) the number of large-scale computer users, and c) the processing capacity of large-scale computers (or "servers"). The growth in fiber optic cable at local telephone companies measures a region's ability to transmit data in digital form, while growth in large-scale computing capacity measures the ability of firms within the region to store and process large databases. As of 1992, the last year of our data, these were important indicators of both advanced computing and telecommunications capital stocks. This capital recently experienced rapid declines in price, increases in quality and rapid expansion nationwide. It has also been associated with transformations in downstream activities and the offering of many new goods and services. We examine the post-divestiture years 1986, 1989, and 1992 because these are years of rapid expansion and also for pragmatic reasons. These were the years for which we could construct a consistent index for all these capital stocks.

We emphasize several findings. First, the level of advanced information infrastructure is unevenly distributed. It locates in especially dense urban areas, rich areas and regions with large amounts of white collar work and related information-intensive industries. Second, advanced information infrastructure grew over time virtually everywhere. In six years the extent of growth has raised the relative standing of all but the worse areas to levels achieved by all but the best (by earlier standards). Third, over time advanced IT infrastructure became less concentrated in any specific region. This holds for the distribution of capital stock and for access to it. Fourth, there has been little shift in the relative standing of different regions, especially at the extremes. Many of the regions that were early leaders still continue to retain their leadership position, and many of the regions that were not leaders still continue to be behind. Fifth, the relative differences among regions have narrowed and many of the regions in the "middle" have changed their relative position.

We conclude that advanced IT did agglomerate in urban areas in any given year but the relative lead of urban areas narrowed over time. Moreover, these data do not support any argument that the distribution of advanced information infrastructure became increasingly more agglomerated in the late 1980s or early 1990s. These patterns are consistent with the diffusion of advanced IT to new uses outside of urban areas. These patterns also support the speculation that conditions were ripe for the rapid diffusion of the Internet. We offer several plausible conjectures about the reasons behind this spread. In addition, we conclude that these data do not support any argument that the diffusion of advanced IT exacerbated regional disparities in economic obtainment in the US.[5]

[5] The US pattern contrasts with other countries. For example, on information technology, see Capello [1994], or Roller and Waverman [2001].

II. Agglomeration and Dispersion in Infrastructure

There is a voluminous literature on the relationship between economic growth and the location and growth of infrastructure in developed countries, where infrastructure refers to the traditional utilities, roads, educational institutions, and so on.[6] There is also a growing literature on the growth of information infrastructure in developed countries, where this refers to telecommunications networks, computing equipment, the Internet, knowledge-intensive industries and related backbone technology.[7] Finally, there is also a somewhat older debate over the appropriate definition of "universal access" in an age of technological change; some aspects of this debate touch on regional disparities in infrastructure.[8] All these discussions would be aided by statistical studies of the regional distribution of the use of advanced information infrastructure.

Below we provide a short summary of the conventional analysis of the forces leading to agglomeration and dispersion of infrastructure. We show how this conventional summary needs to be altered to apply to the spread of advanced information infrastructure. We also briefly discuss why its geographic spread is important for understanding regional economic growth.

1. Advanced Information Infrastructure and Economic Growth

In its most common usage "information infrastructure" is synonymous with communication and computing equipment, software, and related activities associated with information technology. More concretely, this common and broad definition of "information infrastructure" encompasses capital equipment, such as mainframes, minicomputers, PCs, LANs, WANs, local and long distance telephone equipment, private and quasi-public switching equipment and software, both packaged and customized. It also encompasses the results of complementary activities at user enterprises such as training, maintenance and operation. It also involves the flows of tacit and explicit knowledge among enterprises about the technological frontier thorough user groups, trade publications and industry associations.

The term "advanced" – as in "advanced information infrastructure" -- also refers to a similar broad definition, but also to specific recent developments. It partly refers to the use of new technologies in communications networks, where the dramatic leap to digital equipment from legacy analog equipment enabled many new applications and services. In particular, the upgrading of the existing telecommunications networks to digital technology in transmission, switching and other equipment, both in the public switched network and privately operated facilities, accelerated considerably in the 1980s.

The term "advanced" has a slightly different meaning in computing, though it is also associated with new technology. In its common usage "advanced" largely refers to software, peripheral and processing technology near the technical frontier. There was no single

---

[6] See e.g., Munell [1992] or Teubal et al [1996] for a review.

[7] For examples, see National Academy of Engineering [1995], National Information Infrastructure Advisory Council [1995], NTIA [1995], the Information Infrastructure Task Force [1993, 1994], Kahin [1991], Kahin and Keller [1995], National Research Council [1996] and Teske [1995].

[8] For discussions of recent issues and trends, see, for example, Teubal et al [1996], Mueller [1993], Mueller [1997] or Moss and Townsend [1997], Downes and Greenstein [2000].

dramatic technical change comparable to the digital/analog switch in communication; in computing there have been several notable "technical revolutions" (see Friedman and Cornford [1991]). In the 1980s and 1990s these include the PC revolution (associated with thousands of decentralized applications), the departmental computing revolution (associated with super-minicomputers), the on-line-transaction processing revolution (associated with making large databases accessible in real time at large scale computing facilities), the networking revolution (from growth of local area networks to the growth of the Internet), the client/server networking revolution (still on going) and, most recently, the Internet.[9] More generally, the large multiplicity of advances in so many facets of computing creates the perception to users of an ever shifting technical frontier and an ever changing menu of applications.

Broadly speaking, these technical changes bring about economic changes through several related mechanisms. First, there may be declines in the costs of doing an existing activity. This decline in costs should bring about a substitution away from more expensive alternative inputs, expansion in the use of complementary inputs (such as skilled labor), and expansion in enterprises that are information intensive. Second, information technology may enable adopters to perform new activities, potentially associated with the offering of new goods and services. Typically the benefits associated with this type of innovation are realized quite slowly if complementary developments are the least bit complex -- often requiring costly adaptation by using organizations as well as complementary invention by related peripheral and component suppliers.[10] Finally, there is a third mechanism associated with technological competition -- i.e., when many firms innovate in order to match their competitors. When many using organizations adopt an innovation the rents associated with these gains may be competed away. Then the users of goods from information-intensive organizations realize benefits either through decline in prices or increases in new types of applications, which are then passed on to their customers.

All three mechanisms operated in advanced information infrastructure. The change in communication equipment reduced both the costs of maintaining the existing public network and increased the services associated with it.[11] Users also found that they could build greater capabilities into their private communications and computing networks if the local public switch network was of a higher quality.[12] In large scale centralized computing facilities, which are the backbones of most enterprise computing and networking, advanced infrastructure is installed as part of a new system or along with upgrading, retrofitting, and improving existing systems. These changes both reduced the costs of performing existing tasks and enabled the development of entirely new functions.[13]

---

[9] For a history of the structure of the computing industry and its relationship to these technical revolutions, see Bresnahan and Greenstein [1999] and Greenstein [2000].

[10] For a theory of this see Bresnahan and Trajtenberg [1995]. For an application to computing in large organizations, see Bresnahan and Greenstein [1995, 1997] or Brynolfsson and Hitt [1997].

[11] For example, many more services built by local phone companies on top of the existing system -- voice mail, caller-id, remote control of call forwarding, automated paging, alarm monitoring, call-waiting and other enhanced service -- were aided by the advanced infrastructure within the network (Lavoie [1988], Reese [1988]).

[12] For example, higher quality backbone enabled users to invest in less expensive facsimile machines, computer modems, paging services, and other high quality digital communication equipment (Brody[1988], Radford [1989]).

[13] For example, in the 1980s these investments were associated with dramatic improvements in inventory management, decentralization of information applications such as financial analysis and word-processing,

Many of these mechanisms have an important regional dimension to them. For example, the benefits from improvement in local telephone service principally accrue to the local telephone company and local organizations who use these new goods and services. There may also be important regional dimensions to these developments; support services for software and maintenance grow up to service local needs and idiosyncratic problems, contributing to the development of a thick local labor market for advanced IT personnel.

2. The Location of Advanced Information Infrastructure

Advanced IT will tend to agglomerate in urban areas for a variety of reasons. First, in telecommunications networks there are economies of density in the provision and employment of many new services. Second, as long as advanced technology remains relatively unstandardized and difficult to use, the labor market for servicing and operation will tend to favor urban areas.[14] This arises endogenously because it increases the liquidity of labor markets for enterprises that use specialized engineering talent. Similarly, close proximity facilitates learning through user-groups, facilitating the development of complementary service markets for maintenance and engineering services, the delivery of parts and so on.

These agglomeration economies may persist over time for a variety of reasons. Technical frontiers shift frequently, requiring constant experimentation, favoring the users in areas where these experiments are less costly. This would also tend to favor the experienced users in urban areas, and not incidentally, potentially accelerate the relative differences in economic welfare between the leaders and losing areas. More generally, as any of the above factors become even more binding over time, the degree of agglomeration increases.

Agglomeration may not persist as advanced technology becomes less exotic to a greater number of users, which occurs as its features become standardized and its price declines. Then IT diffuses to lower-value uses, to a wider class of users and to many new applications. Many of these new uses and users may not necessarily be located near the first adopters, producing geographic dispersion over time away from the areas of early experimentation. This potentially leads to a decrease in the relative disparity of economic welfare between the late and early adopters. Similarly, as advanced technology becomes less exotic and more standardized, it is also more easily serviced in outlying areas, again contributing to its geographic dispersion.

Political factors may also play a role. In the US, in particular, government policies will mitigate somewhat against the persistence of agglomeration over time, especially when agglomeration results in a notable disparity of economic growth between regions. Competition among localities to attract the key technologies and infrastructure linked to economic growth may lead to direct subsidies of public enterprises or indirect subsidies to enterprises that operate quasi-public infrastructure (e.g., through local public utility policies). There are also national political pressures to reduce regional inequities, especially between urban and rural areas. Several federal agencies have historically pursued policies associated

initiation of automated customer interaction in retailing, and development of applications associated with access to enterprise-wide data and ease of use (Friedman and Cornford [1989]).

[14] For example, labor markets for specialized engineering talent tend to concentrate in a few locations, such as Silicon Valley, Route 128, the research triangle and so on.

with "universal service", which may lead to mandates against regional inequality in access to advanced technology (e.g., apply pressure for the adoption of digital switches in rural areas).

More concretely, several factors arose in the 1980s that could push towards either increasing agglomeration or dispersion of advanced information infrastructure. Here we mention two important ones. First, some industries are more information intensive than others. The geographic dispersion of modern infrastructure will partly depend on whether the information-intensive industries tend to be more geographically concentrated than the less information-intensive industries. Over time changes to this distribution will depend on whether one type of industry tended to grow more in the 1980s and whether the technical changes to information technology induced geographic relocation of information intensive users. Second, networking technology in computing underwent significant changes in the 1980s. For example, the diffusion of on-line transactions processing became linked to the development of many new business services. This encouraged the growth of centralized computing facilities in major enterprises. This could lead to capital deepening at existing users (which may have agglomerated more advanced IT in existing facilities in urban areas) or it could lead to diffusion of more computing locations to new uses in new locations (which may be outside of existing facilities).[15]

While some of these explanations are mutually exclusive, many of them overlap or reinforce one another. Thus, we attempt a modest and feasible research aim which informs understanding in this area. We focus on the following questions:

- What determines the geographic dispersion or agglomeration of advanced information technology across the US?
- How should we characterize the distribution of infrastructure?
- Has the relative degree of disparity between different regions changed in a period of rapid growth in advanced information technology? If so, how?

III. DATA

For purposes of this empirical study we confine attention to a few key components of advanced information infrastructure. We do this for two reasons. First, investment in advanced infrastructure is positively correlated across components within regions. Accurate measurement of a few components gives an excellent indication about the remainder. For example, upgrading to larger CPU capacity on existing computing systems is associated with taking advantage of technical improvements embedded in new systems or taking advantage of technical change in complementary components. Similarly, large investments in digital telecommunications equipment are associated with taking advantage of technical improvements in all digital equipment. Second, there are practical constraints in assembling data. It is simply infeasible to compile comprehensive data of information infrastructure across a wide variety of activities or product markets for the entire US. Accordingly, our

[15] Similarly, the growth of client/server technology encouraged a decentralization of computing activities in the mid 1990s, but extensive experimentation at legacy operations at centralized facilities. While this trend was still incipient at the very end of our sample period (Bresnahan and Greenstein, 1997), it may still have influenced forward-looking investment in the late 1980s.

indices focus on a few components that correlate with advanced technology in digital communications equipment and frontier computing.

Our data set consists of information about local exchange companies (LECs) and large-scale computer users in the U.S. in 1986, 1989, and 1992. It combines data first introduced in Greenstein, McMaster and Spiller [1995] (GMS), in Bresnahan and Greenstein [1995, 1997] (BG), and in Hart, Nave, Raskob and Thomason [1982] (HNRT). Further documentation can be found in the appendices of these papers and of Greenstein, Lizardo and Spiller [1997] (GLS) and Greenstein and Spiller [1997].

1. Sources for Data on Advanced Telecommunications and Computing

Our unit of observation is the territory covered by a local exchange company (LEC) at the sub-state level. This choice is made out of necessity, as no smaller geographic unit is reported to the FCC, and the FCC is the source of our data on telecommunications infrastructure. The shape of these regions within a state depends on how the state divides up local telephone service franchises among different companies. Some states divide up their territory among many companies and some do not divide it at all. Most of these decisions were made years before. The appendix to GLS [1997] includes a list of all these territories, the division of different states, and the major cities within each LEC.

LECs' fiber optic deployment data, as well as demographic and economic data of the counties served by the carriers, come from GMS, while territorial extension data are from GMS and HNRT.[16] There are 101 LECs that annually report their fiber deployment to the FCC.[17] Demographic, economic, and territorial data of the counties served by each LEC, which are used in the regressions, are aggregated to the company level within each state.[18] While some small independent companies do not report to the FCC, the GMS data cover virtually all of the fiber optic cable used by local telephone companies.[19] Figure 1 shows a visual approximation of the geographic area corresponding to the 101 LECs reporting to the FCC (it is approximated at the county level, not the LEC territory level). This area covers almost all the U.S. territory and over 90% of the US population.

---

[16] The land areas from HNRT are for LEC square mileage in 1981. There may have been slight changes over time due to mergers and other transactions.

[17] If a LEC generates more than 100 million dollars in revenue a year, it must report its infrastructure data to the FCC. Fiber deployment is reported using M and ARMIS 43-07 forms. The total revenue associated with the local telephone firms deploying the fiber in our index exceeds seventy billion dollars in 1992.

[18] The sources of the demographic data are the "Annual Estimates" of the U.S. Bureau of Economic Analysis, Department of Commerce, and "Regional Economic Information System Annual CD." The information is gathered for 48 states and the District of Columbia. Alaska and Hawaii are excluded. See GMS for detail.

[19] It includes the vast majority of cable at local telephone companies in the US. Many of these areas are geographically small or, by definition, do not concern populous areas. In addition, this data does not include investment in fiber optics by competitive access providers such as Teleport or MFS, not does it include rural independent LECS, neither of whom are required to report their capital stocks to the FCC. As of 1992, this was not a large percentage of any region's fiber network. Including these providers might be an important component of a region's fiber if this data were updated to the present.

Figure 1. Operating Area 101

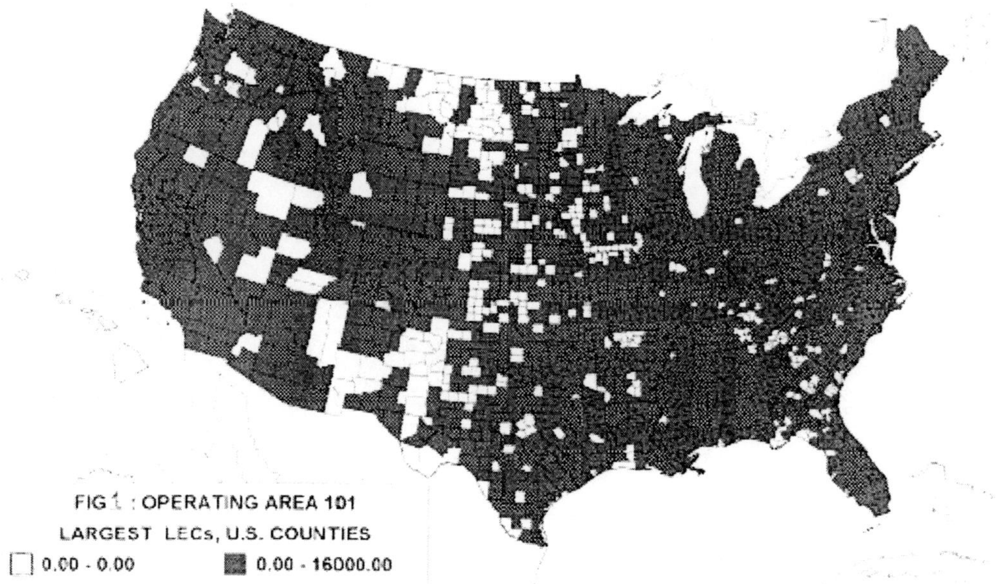

Information about large-scale computer users' locations and computing capacities overlaps with that found in BG. These data are collected by the Computer Intelligence Infocorp (CII), a division of Harte Hankes, a market survey firm based in La Jolla, California, that regularly calls computer users in businesses to document their computer use patterns. We use the survey results from the end of each calendar year. The magnitude of the sample of large-scale systems users in each year is 13,788 sites in 1986, 13,553 sites in 1989, and 12,386 sites in 1992. Each of these users had at least one mainframe or large supermini computer at their site when they were surveyed. CII claims to cover 70% of all large-scale users in the U.S. These users cut across the entire spectrum of US industries, with some slightly better coverage of the Fortune 500 and facilities of major corporate computing facilities.[20] Since there is also extensive coverage of thousands of firms outside this group, there seems to be no reason to expect the error in CII's sampling of large-scale users to correlate with the geographic location of users.

We matched the CII data against the FCC data through the use of zip codes (primarily), city codes (secondarily), and county codes when available (rarely), relying on GMS when no other information was available. We have found that virtually all the large-scale systems users are located in the geographic areas served by the 101 largest LECs that report fiber optic deployment to the FCC. In each year, less than 4.5% of sites included in the CII sample belong to territories that are served by LECs which do not file with the FCC. Further, the proportion of overall computer capacity, as defined below, located in these territories is less than 2.7% of the total amount in each year. Thus, except in Table 5, all our analysis is based on information for those LECs that report annually to the FCC and for the computer users

[20] BG [1996] found that for sites that contained identifiers approximately 15% of the observations came from Fortune 500 firms and that over 85% of the Fortune 500 have at least one representative in this sample.

within those LECs. Figure 2 shows the geographic distribution of computer users relative to our coverage of LECs, visually approximated in the same way as Figure 1.

Figure 2. Sample Coverage by U.S. Counties, 1992

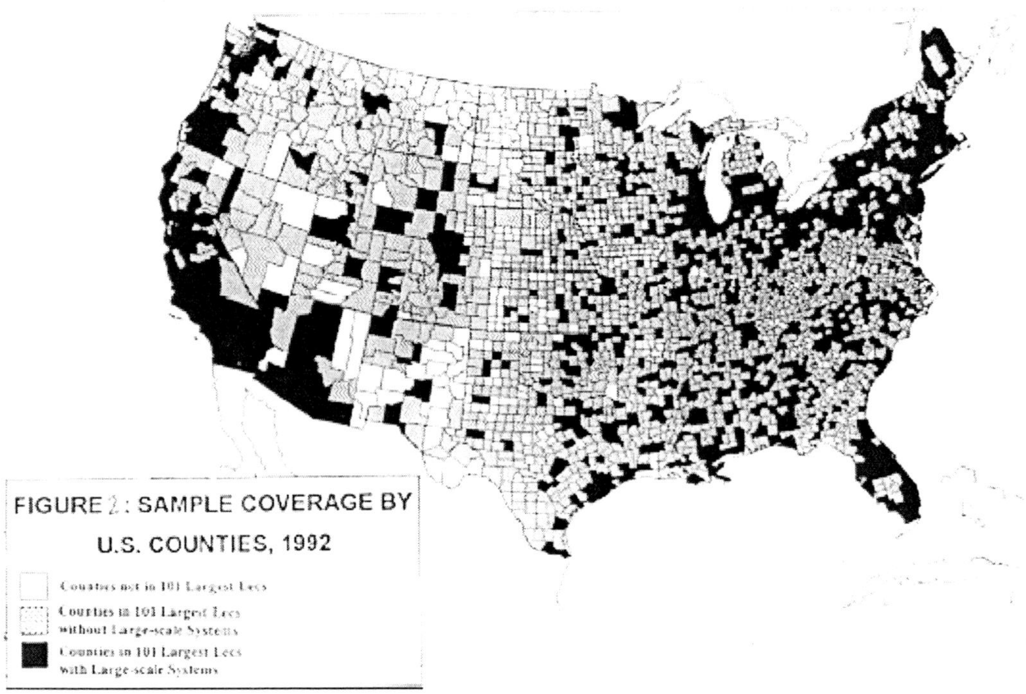

2. Indicators of Advanced Information Infrastructure

We use three indicators of the stock of information technology infrastructure: the number of sites using large-scale computing, their associated computing capacity, and the miles of fiber optic cable used by the LEC within a region. We now explain these choices.

The number of sites using large-scale computing is one simple indicator of the presence of sophisticated users in a region. Another related indicator of infrastructure is the computing capacity existing within a region, which we measure by the sum of Mips (millions of instructions per second) of all processors within computers systems existing in a region's sites. We also include all the client/server hardware investment at these sites, reflecting the spread of the networking revolution to large-scale computing in the 1990s. Precise definitions for the computing models included and excluded may be found in BG [1997].[21]

In the early years of our sample either of these indices measures the bulk of large-scale computing capacity in a region. BG [1996] and [1997] show that the vast majority of these

[21] This includes all mainframe and non-mainframe computing hardware except PCs at all the sites operating within the region boundaries. We found in preliminary analysis that there was a strong correlation between aggregate mainframe MIPS and aggregate non-mainframe MIPS within a LEC in a year. Thus, for the purposes of this study, there was no benefit to distinguishing between the two different types of capacity or weighting them differently. We present the unweighted total.

sites are associated with white collar administrative work at central computing facilities. A smaller, and slightly declining, fraction is associated with engineering or manufacturing work. Thus, the only systematic bias is associated with sites which left the sample. BG [1996] show that the diffusion of client/server networking up until 1992 primarily occurred at engineering oriented sites and that most non-random exits, which were still small as of 1992, were concentrated at these engineering sites.[22] Hence, our measure of a region's computing capacity slightly emphasizes computing at administrative sites in 1992; We think this emphasis will be small since we also add non-mainframe MIPS -- i.e. the MIPS associated with servers in client/server configurations -- into our total measure.

The presence of a site or the extent of MIPS in a region indicates some significant developments of advanced IT. There is no way for most of these facilities to operate smoothly without the presence of many complementary technologies, without users with technical skills, nor without the presence of a support organization or other related infrastructure. These are simply too big an investment to live in isolation of other technical advances. Large scale systems cost approximately half-a-million dollars for the hardware on average. At most large facilities, the software, training, installation and maintenance expenses are three to four times the hardware expenses on an annualized basis (BG).[23]

Our last indicator is the number of miles of fiber cable installed by local telephone companies within a region, where "miles" has been adjusted to reflect the number of strands of fiber. Fiber optic cable enables high-volume transmission of data and is widely considered to be an essential part of the modern digital telephone system. The limits on its capacity are determined by the available terminal and repeater technology. Due to all the complementary factors that must be in place to make it effective, this too is an excellent measure of advanced communication infrastructure. Further digital technologies, deployed in conjunction with fiber, enable a variety of services. Fiber enhances the reliability as well as the quality of data services over phone lines. (See, e.g., Brody [1988], Lavoie [1988], Minoli [1991], chapter 7, Reese [1988]).

Although the miles of fiber optic cable used by the LEC within a region includes neither fiber deployed by long distance telephone companies nor that deployed by private providers and users of local fiber optic cable, our measure reflects the bulk of fiber devoted to local telephone networks and/or data transfer over the public network phone lines. It reflects tens of billions of dollars invested in capital stock using digital technology. Much of this investment received publicity when announced and may have been the focus of negotiation between local regulators and utilities (see GMS [1995]).[24]

---

[22] As with the diffusion of most advanced technologies, there was resistence among users of legacy technologies to completely ceasing use of the old systems. There were many reasons for this, including worries about losing site-specific investments, uncertainty over future technologies, and so on. As a result, the demonstration of many working prototype client/server systems in 1992 did not induce much immediate retirement of legacy systems. See Bresnahan and Greenstein [1997, 1997].

[23] Spread over 12,000 sites in 1992 and more sites in the earlier years, the total value of the capital going into the computing site of our index must be in the tens of billions of dollars. As shown in BG [1997], the value of hardware associated with large scale systems continued to grow into the 1990s. Its fraction of the total computing capital stock did not dramatically change until the mid 1990s.

[24] We are also leaving out investment in fiber optics by small rural telephone exchanges, most of whom are located in small cities and towns scattered throughout the US. To the extent that this investment represents real economic advance and not just gold-plating investment under regulatory distortions, it will only enhance our findings below that there has been a spread of fiber from urban to rural areas. However, many of these small cities will not have large investment in large scale computing, so we think it is unlikely that our measurement

3. Measuring Access to Infrastructure

To measure the level of access to advanced information technologies we borrow from the established methods for measuring access to infrastructure in developing countries (Cowell [1995]). In this literature researchers examine changes in the "density" of infrastructure. We experimented with a variety of methods for measuring "density." As it turned out, all our experiments pointed toward the same qualitative answer, so the composite measure will be the best single index for a region's standing. These indices are:

- *Number of Sites/Population*[25] measures whether a worker living in a region with a high number of Sites/Population ratio has more chances to be exposed to computer technology.
- *All Systems Mips/Population* measures the exposure to computing power of large-scale systems existing in the region.
- *Fiber/Land,* which equals Miles of Fiber Optic Cable Deployed/Miles of Land Extension, measures the deployment of fiber optic across geographic regions. A LEC with a high Fiber/Land ratio has a greater capacity to supply advanced telecommunication services than one with a low ratio.[26]
- *Networking Capacity Index*, which equals (Fiber/Land)*(MIPS/Population), is a composite index that measures the potential capacity of a geographic area for storing information and transmitting it among computers in the region using the public telecommunication infrastructure provided by the local exchange company.[27] It is analogous to transportation infrastructure indices of roadway and vehicle carrying capacity per capita and per square mile.

In our exploratory research of these data (GLS [1997]) we found that all of these measures of access are highly correlated on a regional basis. Thus, in general, we will focus primarily on one summary measure, the networking capacity index.[28]

4. Infrastructure Statistics: An Overview

The geographic distribution of advanced large-scale computer and fiber infrastructures in the U.S. is deceptively skewed. Many geographic areas contain few large-scale computer

---

framework and qualitative results would be greatly influenced by their inclusion. We await further evidence of diffusion of advanced IT to small rural towns before reaching a definitive conclusion.

[25] The numerator in the indices Number of Sites/Population and All Systems Mips/Population corresponds to 1,000 people.

[26] For this calculation we use the square miles of the LEC within the state relative to its fiber from HNRT.

[27] A region without fiber optic deployed, but with a high processing capacity of information, has a Networking Capacity Index of zero. So too does a region with fiber and no computing.

[28] We also explored another standard composite index used by economic geographers to measure the amenities associated with different locations (See Nissan [1994] for a full explanation). This index combines the information provided by the first three aforementioned sub-indices, Number of Sites/Population, All Systems Mips/Population, andFiber/Land. It normalizes each sub-index and then for each region sums a weighted index of the difference between the maximum value reached in the sample and each normalized value in any given year. The larger the distance from the maximum the smaller the index. It is well known that this index is inappropriate with skewed data. Since the distribution of infrastructure is highly skewed, this procedure is flawed here. See GLS [1997] for these calculations.

users, low or no computing capacity, and low or no deployment of fiber optic cable in relation to geographic areas in the top of the distribution. However, most of the U.S. population lives within the territories that are advanced in their IT deployment. This is because most of the US population is located in a small number of regions, most of which are advanced. Table 1 contains the descriptive statistics for the distribution of advanced information infrastructure in the late 1980s and early 1990s.

Table 1: Descriptive Statistics of Computer and Fiber Optic Infrastructures LEC Level

Descriptive	No. Sites			All Systems MIPs			Fiber Optic (Miles)		
Statistics	1986	1989	1992	1986	1989	1992	1986	1989	1992
Total	13257	12988	11836	112,476.2	258,666.6	553,093.5	265472	2443449	6316436
MEAN	131.3	128.6	117.2	1,113.6	2,561.1	5,476.2	2,628.4	24,192.6	62,539.0
MEDIAN	47	48	41	270.2	678.1	1,373.4	151	9138	22371
STD	218.2	208.9	184.9	2,030.7	4,477.0	9,497.3	7,148.3	37,972.0	89,721.6
CV	166.2	162.4	157.8	182.3	174.8	173.4	272.0	157.0	143.5
MIN	0	0	0	0	0	0	0	0	0
MAX	1291	1261	1071	12,980.4	27,042.2	59,630.1	50553	203561	451356

In Table 1 we see that our three indicators of the stock of information infrastructure experienced changes in the distribution during the period 1986-1992. The number of sites that used large-scale computers decreased during the period, while the computer processing capacity and the deployment of fiber optic grew remarkably fast. The total number of millions of instructions per second (All Systems Mips) was almost five times larger in 1992 than in 1986. The miles of fiber optic deployed by the 101 largest LECs was nearly 24 times larger in 1992 than in 1986. The situation in 1986, when around 28% of the 101 LECs did not have fiber optic deployed in their operating areas, contrasts dramatically with the situation in 1992, when just around 4% did not. This advanced IT capital has diffused at an extraordinarily rapid rate. The only comparable historical precedents for this kind of growth are the well-known massive efforts behind the building of other communication or transportation infrastructure -- i.e., the building of miles of post-civil war rail, pre-civil war telegraph lines, turn-of-the-century telephone and (slightly later) electrical lines, and miles of post world-war II national highways.

The percentage of population living in regions with relatively low stock of information infrastructure is small. Less than 15% of the population of the 101 largest LECs are served by LECs that have less than the median Number of Sites, All Systems Mips and Fiber Optic. In the case of fiber, the fraction of population that is served by the LECs that belong to the top ten percentile has increased from 15.4% in 1986 to 44.25% in 1992.

In the descriptive statistics we see hints that the geographic dispersion of the three variables increased during 1986-1992. The coefficient of variation of the geographic distribution of large-scale computer users, Mips and Fiber Optic is consistently lower across the years. The largest changes in the coefficient of variation of the geographic distributions of Mips and Fiber Optic occurred during the period 1986-1989.

In Table 2 we present the descriptive statistics of access indices in two different forms. One includes Washington D.C. in the sample and one does not.[29] Clearly, there is variance in the level of access to advanced IT across the U.S. While it is difficult to evaluate whether these differences are large or small, as there is no natural baseline against which to compare them, the degree of access increases over the time in three of the indices -- fiber/mile, MIPS/pop, and the networking index. We note that there is little to be inferred from declines in site/pop, since the total number of sites is declining nationwide, potentially due to the client/server revolution, while the total number of MIPS/pop grows dramatically.

Table 2: Indices of Access to Computer and Fiber Optic Infrastructures Descriptive Statistics: LEC Level

	Fiber Optic /Land			All Systems MIPs /Pop			Number of Sites /Pop			Net-Working Index		
	1986	**1989**	**1992**	**1986**	**1989**	**1992**	**1986**	**1989**	**1992**	**1986**	**1989**	**1992**
Washington D.C. Included												
MEAN	0.153	3.363	7.628	0.408	1.007	2.066	0.059	0.059	0.052	0.117	10.969	43.317
STD	0.424	14.942	31.725	0.424	1.328	2.259	0.102	0.102	0.078	0.552	79.895	298.049
CV	277.700	444.400	415.900	103.750	131.800	109.330	172.400	173.980	151.400	470.244	728.365	688.068
MIN	0	0	0	0	0	0	0	0	0	0	0	0
MAX	2.564	148.800	312.522	2.593	10.782	15.130	1.026	1.256	0.779	5.240	800.180	2983.170
Washington D.C. Excluded												
MEAN	0.134	1.908	4.579	0.386	0.964	1.991	0.058	0.057	0.050	0.066	3.077	13.918
STD	0.382	3.119	8.262	0.364	1.259	2.141	0.101	0.101	0.077	0.201	9.663	39.461
CV	285.190	163.459	180.440	94.202	130.634	107.511	175.855	177.516	153.734	304.024	314.047	283.522
MIN	0	0	0	0	0	0	0	0	0	0	0	0
MAX	2.564	15.601	63.143	2.577	10.782	15.130	1.026	1.256	0.779	1.274	85.934	326.132

Finally, Figure 3 summarizes the network capacity index in a picture. It classifies the LECs' geographic areas by the level of their networking indices in 1986, 1989, and 1992. The darkest group corresponds to those areas with a networking index in the top decile of the distribution for 1989. The second darkest group (dark grey) represents the areas with a networking index above the median but below the top decile in 1989. The next lightest shade (of gray) is for those areas with a networking index above zero and below the median in 1989. Areas with nothing are white.[30] We apply the same cardinal levels for determining the shades to the maps for 1986 and 1992.

[29] Washington D.C. is the only major urban hub of such a small geographic size to be so dense, to have its own telephone regulator and report separate statistics to the FCC. Washington D.C. is a dense urban area comprised of much administrative work and low residential population and small geographic area. As a result, it reports extraordinarily high per/capita and per/mile statistics for all our infrastructure data. This one observation strongly skews the distribution of any access statistic we construct. We are fairly certain that similar statistics would result if we could collect them for a comparable area of downtown Manhattan, Boston's financial district, Chicago's business loop, downtown San Francisco, or any other major urban hub. However, all these other urban areas are combined with much of the residential outlying regions to form the LEC territory.

[30] Though our rankings come from networking indices computed at the LEC level, we white out all counties where we do not find any large-scale computer users (even if the local telephone company primarily responsible for

Figure 3. Networking Index by U.S. Counties

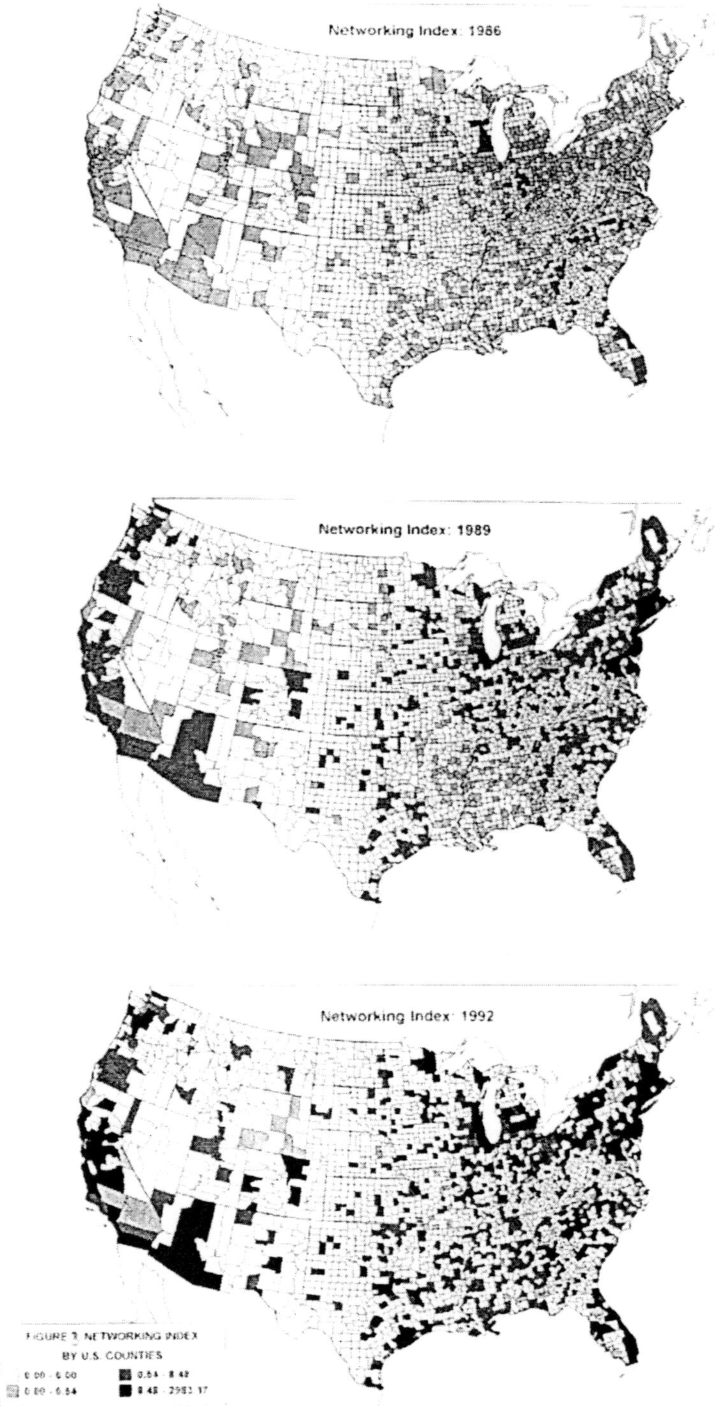

that county has fiber optics in other parts of itsregion). This is an approximation and, yet, a more accurate visualization of the distribution of access since it also reflects the distribution of industry and population. Fiber optics in the local exchange are unlikely to be located where there is no industry.

As we can see from the maps, in any given year, advanced IT agglomerates in a few regions. The highest levels of information infrastructure are located in the big urban centers of the U.S. No single region or small number of cities, with the possible exception of Washington DC (as noted above), dominates the country. Overall, the distribution follows the U.S. population, and it appears from the picture that the major urban centers have the most advanced infrastructure. In 1992, the top areas include LECs that cover San Francisco, San Diego, and Los Angeles, Seattle, Tucson and Phoenix, Denver, Houston, Dallas, Minneapolis, Chicago, St. Louis, Detroit, Indianapolis, Cincinnati and Cleveland, Pittsburgh and Philadelphia, Buffalo, New York City and environs in Connecticut and New Jersey, Boston, Washington D.C. and its environs, Raleigh-Durham, Atlanta, Miami, St. Petersburg and Tampa.

Second, much has changed over time as there is investment in advanced IT infrastructure. All three maps together illustrate growth over time (as they get darker over more and more regions). In 1986 only six LECs had a networking index above the 1989 median. In contrast, in 1992 there were 33 LECs with a networking index above the 1989 top decile. Looking at the evolution over time, differences between regions never completely disappear. Yet, we also observe changes in those areas that occupy the different ranks.

IV. Explaining the Cross Sectional Distribution

The location of information infrastructure in a region depends on the supply and demand for investment, which depends on particular characteristics of the region and many other complex factors. Below we offer a summary of the factors that influence the agglomeration of infrastructure in some regions and not others. A simple regression describes the cross-sectional pattern of information infrastructure during 1986, 1989, and 1992. Our main point is that the density of human settlement and the location of industry alone explain much of the geographic pattern of advanced information infrastructure, though certainly not all of it. In addition, we also show that the pattern of IT capital is less concentrated in regions specialized in information technology intensive activities than in regions specialized in less IT intensive activities. This suggests that dispersion of IT-using industries contributed to increasing the dispersion of advanced IT in the 1990s.

1. A Descriptive Regression

We estimate a reduced form equation for the networking index existing in a region as a function of a set of demographic and economic variables of the region. The data correspond to the 101 local exchange companies over the years 1986, 1989, and 1992.

We explain advanced information infrastructure by the local per capita income, population density, the fraction of the local employment devoted to finance, insurance, and real state (FIRE), the size of urbanized and rural populations, and a set of dummy variables that describe the characteristics of the region's main cities.[31]

[31] An explanation for the construction of these variables can be found in Greenstein and Spiller [1997].

Local per capita income directly measures the earnings of the local population, but it is also correlated with other factors, such as tax revenues, high-value-added businesses in the region and so on. Population density, urbanized population, and rural population are key factors in the decision to deploy fiber optic cable and advanced computing. Urbanized population represents the population served by the LEC in cities with a population of 50,000 or more, while rural population represents rural settlements with less than 5,000 people. The fraction of local employment in FIRE identifies the importance in the local economy of a group of activities intensive in data processing and transmission. All these variables, except rural population, should have a positive impact on the location of information infrastructure.

We also include a set of dummies to identify the characteristics of the cities included within the boundaries of the local exchange company. One of these dummy variables measures whether the LEC contains within its boundary a city with a population over a quarter million inhabitants in 1990. This variable captures the existence of a critical mass of potential business users of information infrastructure located in an urban hub. Another dummy variable specifies whether the LEC covers an area considered (by the U.S. census) as one of the fifty fastest growing areas in the U.S. between 1980 and 1990. Most of the fast growing areas in the 1980s were extensions of suburban communities, rather than concentrated urban hubs; this variable could predict under-deployment of advanced information technologies or the opposite, depending on whether recent growth makes deployment of advanced information technology easier or more difficult.

Descriptive statistics for the variables in the regression are presented in Table 3, and the regression results are in Table 4. A test of structural change shows that we can reject the hypothesis that the regression parameters are the same not only in the three years taken together, but also in either of the sub-samples of 1986-1989, 1986-1992, and 1989-1992. Thus, we report a separate regression equation for each year. In these the explanatory power of the regression increases over time (the R-squared increases), suggesting that the distribution becomes easier to predict as cumulative investment resulted in large unchanging capital stocks in many places. Yet, the elasticities of the important variables also change from one year to another, suggesting overall that the relationship between advanced infrastructure and exogenous factors is changing over time.

We see that density predicts advanced infrastructure location. In 1986, a 10% increase in population density led to a 0.8% increase in the networking index, while in 1992 it led to 3.8%. Variation in population density across regions induces substantial differences in the networking indices prevailing in those regions. In 1986, when evaluating the impact of other regressors at their mean level, one standard deviation around the mean of log of population density showed that the networking index fluctuated from 0 to 0.20. During that year approximately 90% of the sample had a networking index in this range. A similar calculation at the mean of log of population density in 1992 would have created a fluctuation in the networking index from 2.362 to 7.939. This interval includes the middle 20% of the sample in 1992.

Infrastructure also appears to be a complex function of density. While the fraction of urban population in a county increases the amount of infrastructure, the presence of a city with population over a quarter million within the boundaries of an LEC makes a difference in the level of the networking index too. Holding constant the influence of other variables, the presence of a relatively large city increases the expected value of the networking index by 65%. The variable FASTGROW decreases the expected value of the networking index by

63% in 1992. These results suggest that the largest agglomerations of advanced infrastructure are found in dense urban areas with established cities in the US -- the traditional financial centers -- not those growing rapidly with suburban sprawl.

Table 3: Descriptive Statistics Endogenous and Exogenous Variables

	MEAN	STD. DEV	MEDIAN	MIN	MAX	NUM. OBS
ENDOGENOUS VARIABLE						
LOG (1+NETWORKING INDEX)						
1986	0.070	0.226	0.003	0.000	1.831	101
1989	0.781	1.089	0.348	0.000	6.686	101
1992	1.612	1.489	1.322	0.000	8.001	101
EXOGENOUS VARIABLES						
LOG PERCAPITA INCOME						
1986	9.544	0.155	9.537	9.229	9.919	101
1989	9.600	0.161	9.579	9.283	10.027	101
1992	9.630	0.147	9.616	9.298	10.045	101
LOG POPULATION DENSITY						
1986	4.460	1.303	4.397	1.442	9.249	101
1989	4.482	1.310	4.439	1.442	9.227	101
1992	4.515	1.305	4.480	1.429	9.162	101
FRAC POP EMPLOYED IN FIRE						
1986	0.067	0.017	0.067	0.034	0.117	101
1989	0.066	0.018	0.065	0.034	0.118	101
1992	0.064	0.017	0.062	0.032	0.116	101
CITY OF HALF MILLION	0.178	0.385	0.000	0.000	1.000	303
CITY OF QUARTER MILLION	0.376	0.487	0.000	0.000	1.000	303
FAST GROWING AREA	0.149	0.357	0.000	0.000	1.000	303

Table 4: Log Networking Index OLS Regression Results

Regressors	All Years	1986-1989	1989-1992	1986	1989	1992
Dummy 1986	-7.399 -(1.348)	-5.504 -(1.245)		1.572 (0.965)		
Dummy 1989	-6.710 -(1.216)	-4.815 -(1.089)	-11.926 ** -(1.760)		-10.393 -(1.557)	
Dummy 1992	-5.879 -(1.064)		-11.104 -(1.635)			-14.914 -(1.293)
LNPERCAP	0.506 (0.846)	0.409 (0.858)	0.952 (1.294)	-0.190 -(1.085)	0.861 (1.192)	1.293 (1.038)
HALFMIL	-0.058 -(0.322)	0.008 (0.004)	-0.133 -(0.636)	0.117 (1.516)	-0.094 -(0.365)	-0.173 -(0.562)
QUARTMIL	0.272 * (2.009)	0.127 (0.957)	0.443 * (2.522)	0.018 (0.530)	0.287 (1.341)	0.603 * (2.235)
FASTGROW	-0.283 * -(2.603)	-0.161 -(1.629)	-0.408 * -(3.141)	0.027 (0.487)	0.324 -(2.376)	-0.475 * -(2.334)
FIREEMP	13.485 * (2.430)	6.554 (1.386)	17.800 * (2.549)	-0.448 -(0.288)	11.741 (1.618) *	22.630 ** (1.975)
LNPOPDEN	0.379 * (6.940)	0.272 * (4.487)	0.518 * (8.035)	0.070 ** (1.850)	0.467 * (5.117)	0.557 * (6.292)
Adj. R-squared	59.485	46.662	66.872	17.300	61.452	66.783
Number Observed	303	202	202	101	101	101
Residual Sum of Squares	183.652	76.789	119.919	3.970	42.934	69.237

Numbers in parentheses are t-statistics. All standard errors and associated t-statistics in the table were computed using White's covariance matrix estimator and are heteroskedastic-consistent.

* Significant at 5%

** Significant at 10%

We see evidence that industry and income matter. This is provided by the importance of FIRE. A larger fraction of employment in FIRE represents a large financial and white-collar sector. This increases the information infrastructure index. Broadly interpreted, it is not surprising that administrative work and infrastructure move together, but it is difficult to infer causality from this in the long run. Also, per capita income becomes important in the last two years of the sample. The implied elasticity on income in 1992, which is 1.9, suggests that rich regions tend to have more infrastructure.

Finally, we note a puzzling pattern over time. In the restricted regression, where all coefficients except the time dummies are set equal, the implies rates of growth are enormous. The difference in the time dummies imply an average of 70 percent growth in information infrastructure across all the observations between 1986 and 1989, holding constant for other factors. Similarly, we see a 100% growth between 1989 and 1992. Yet, when the coefficients are allowed to differ across years, much of these differences over time show up as differences in the relationship between infrastructure and its predictors. This suggests that the distribution of infrastructure may be changing over time as the relationship between infrastructure and exogenous factors change. These changes may reflect a resettling of post-divestiture investment patterns in telecommunications and a change in computing investment in response to the networking revolution. Since several factors simultaneously influence this result, further work will need to disentangle its components.

In sum, densely populated areas tend to have more access to infrastructure, with the most densely populated cities having the most advanced infrastructure. Yet, agglomeration reflects more than just density. It helps to be a rich area, as well as one with a larger white-collar sector, and it does not help to be one experiencing recent rapid growth in population. In addition to these cross-sectional features, there is more advanced infrastructure everywhere over time, a trend which resulted in virtually all areas deepening their advanced information capital at extraordinarily rapid rates.[32] The relationship between exogenous factors and advanced IT also seems to be changing over time; this motivates further analysis of the changes to the distribution.

2. Location of Information Technology Intensive Economic Activities

In this section, we take advantage of additional information found in the computer user surveys to characterize the role of white-collar activity in determining the location of information infrastructure. CII reports the primary industry for each computer using site. We use this information to analyze the contribution of the distribution of different information-intensive activities to the observed distribution of infrastructure.

Table 5 shows that the economic activities with the larger number of sites using large-scale computers were, in 1992, business services, wholesale trade, education services, depositary institutions, insurance carriers, health services, and computers and related equipment.[33] These activities accounted for 46% of the total number of sites. The question we investigate is this: if these intensive users of information technology did not alter their location in response to more investment in advanced IT, would more investment lead to more or less agglomeration of advanced IT? We can get some evidence by looking at the geographic distribution of information-intensive users.

---

[32] Because of the skewness of the data and the presence of outliers, we were concerned that OLS does not properly model the determinants of infrastructure. In Greenstein and Lizardo [1997] we perform quantile regressions (Koenker and Bassett [1978]) that allow us to estimate different points of the conditional distribution of the networking index. Our inferences did not substantially change, so we only report the OLS results here.

[33] This excludes federal/state/local government.

Table 5: Characteristics of the Distribution of All Systems MIPs by Group of Economic Activities, 1992

Economic Activities	SIC Groups	Econ. Activ. Share All Sys MIPs	Num LECs w/Bus in Econ. Activ.	GINI Coefficient Distrib. of Capacity Specializ. Index
Federal/State/Local Government	91-97	14.85	85	0.593
Business Services	73	11.93	79	0.604
Wholesale Trade	50-51	2.64	73	0.695
Education Services	82	6.63	83	0.647
Depositary Institutions	60	5.30	81	0.716
Insurance Carriers	63	7.95	66	0.618
Health Services	80	4.09	77	0.638
Computers and Related	35, 365-368	5.44	67	0.720
Legal, Social, Engin. Serv., Museums	81,83,84-89	3.50	55	0.764
Misc. Retail	52,55-59	1.83	58	0.782
Printing and Publishing	27	1.46	57	0.717
Manufacturing	21-25, 29	1.74	59	0.871
Chemicals	28	2.38	44	0.862
Misc. Manufacturing	30-32, 39	0.91	60	0.865
Transportation Services	40-47	2.99	47	0.785
Insurance Brokers and Real Estate	64-69	2.27	48	0.787
Fabricated Metals	34	0.46	49	0.831
Electrical Apparatus	361-364, 369	1.08	45	0.894
Food Products	20	0.96	48	0.909
Gas and Sanitary Services	492-495	1.61	52	0.796
Other Transportation Equip.	372-379	2.67	46	0.858
Instruments	38	0.95	42	0.872
Motor Vehicles and Equip.	371	1.31	39	0.876
Primary Metals	33	0.52	46	0.934
Security and Commercial Brokers	62	1.94	27	0.891
Electric Services	491	1.49	55	0.785
Food Stores	54	0.49	40	0.863
General Merchandise Stores	53	1.71	42	0.883
Non-Depositary Institutions	61	1.24	43	0.852
Paper and Allied Products	26	0.57	42	0.847
Telephone Communication	481	3.09	44	0.765
Hotel, Personal Services	70-72,75-79	0.62	30	0.903
Mining and Construction	10-12, 14-17	0.64	37	0.922
Other Communication Services	482-489	0.31	28	0.906
Oil and Gas Extraction	13	2.35	13	0.950
Agric, Forestry, Fish., and Hunting	1,2,7,8,9	0.08	17	0.964
Descriptive Statistics				
Mean		2.78	50.67	0.810
Std. Dev.		3.19	17.65	0.103
Median		1.72	47.50	0.850
Min		0.08	13.00	0.593
Max		14.85	85.00	0.964

Economic Activities are presented in descending order of Share of Sites.

$$\text{Index} = \frac{\text{Mips in Region i used in Activity j/Mips in Region i}}{\text{National Mips used in Activity j/ National Mips}}$$

We employ an index of geographic specialization in computer intensive economic activities that follows Krugman [1991]. We construct an index that measures the degree of specialization of the computing capacity existing in a region in a particular economic activity. For each group of economic activities we construct the index as follows:

> If, for a given economic activity j, a region shows a ratio greater than one, we say that the computing capacity existing in this region is more specialized in the generation of activity j than is the computing capacity existing in the rest of the regions that present a ratio less or equal to one.

Table 5 shows a high inequality of computing capacity across information-intensive sites relative to less information technology intensive economic activities. The distribution of the computing capacity across regions is relatively less concentrated in the former group of activities than in the latter. Moreover, economic activities included in the former group are developed in a much larger number of regions.

For those activities less intensively using information technology, the geographic distribution of the specialization index of computer capacity looks like the locational pattern of industrial employment discussed in Krugman. Many industries in the U.S. are highly geographically concentrated and most of these industries are not high tech sectors. Table 5 suggests, therefore, that the agglomeration (or lack thereof) of advanced information infrastructure is partially a function of the growth of information intensive industries, which, themselves, are more dispersed than the average industry.

We must state these conclusions carefully. At most we can say that the geographic distribution of white-collar industries contributes to a greater spread of advanced information infrastructure than would be typically found for an important input in non-information intensive industries, such as manufacturing. As a greater fraction of the economy moves into these information-intensive industries, this movement could contribute to the spread of information infrastructure. As yet, this evidence does not allow us to assign any causality. Does the cheaper input cause the information intensive activities to expand or does the expansion of these industries cause the spread of advanced IT? These are simultaneous events and the present methods can only allow us to make the association between them. To make further inferences about the contribution of white-collar work to the geographic diffusion of advanced information technology, we need further evidence on the changing geographic distribution of information-intensive work over time.

V. Changes to the Distribution of Infrastructure

We now summarize changes to the distribution of advanced information infrastructure. We use a standard set of measures found in the study of income inequality and the study of infrastructure in developing countries: quantile ratios, the decile distribution ratio, and the Gini coefficient. We use all three to show that our results are not sensitive to method; using all three we find that advanced IT became less agglomerated over this period.

The analysis of quantile ratios allows us to compare the proportionate movement of the different quantiles of the distribution of the variable of interest. The quantile ratio is defined as the ratio Qth-quantile/Median. The comparison over time of these ratios illustrates whether the quantiles are moving closer to each other or if they are moving apart over time. A simpler and related measure is the Decile Distribution Ratio (DDR), which is the ratio of the share of the bottom 40% in relation to the share of the top 20%. When inequality decreases, this ratio increases. Finally, the Gini coefficient is defined as:

$$G = \frac{1}{2n^2 \bar{y}} \sum_{i=1}^{n} \sum_{j=1}^{n} | y_i - y_j |$$

where y_i and y_j are the values taken by the variable in regions i and j, and n is the total number of regions. The definition corresponds to the average difference between all possible pairs of the variable among the regions, expressed as a proportion of the total sum of the variable. The Gini coefficient takes values in the interval [0,1], 0 representing total equality, and 1 representing the maximal inequality.

1. Changes to Information Infrastructure

The quantile ratios (Table 6) show that the information infrastructure rich do not seem to be getting richer, and the poor are improving their relative position. The bottom halves of the geographic distributions of MIPs, Sites and Fiber are getting closer to their medians. In the case of MIPs and Sites, there is no clear tendency in the movement of the quantile in the upper half of the distribution. In the case of Fiber, there is a tendency toward equalization of the amount of fiber deployed by the upper half of the distribution in relation to the median.

Table 6: Quantiles and Quantile Ratios Geographic Distribution of Computer and Fiber Optic Infrastructures

	No. Sites			All Systems MIPs			Fiber Optic (Miles)		
	1986	**1989**	**1992**	**1986**	**1989**	**1992**	**1986**	**1989**	**1992**
Quantiles									
10th	2	2	2	6.2	14.0	61.9	0	487	1,397
20th	13	13	13	52.3	109.9	274.5	0	2,053	5,442
30th	20	23	22	116.7	254.4	637.3	13	3,248	8,227
40th	32	34	30	206.9	471.6	963.4	89	4,830	15,344
50th	47	48	41	270.2	678.1	1,417.7	151	9,138	22,371
60th	69	63	60	492.8	1,261.9	3,023.2	365	14,798	42,111
70th	114	116	108	874.4	1,942.6	4,506.8	557	23,611	62,067
80th	197	194	173	1,774.9	3,536.4	7,931.6	1,390	33,848	93,111
90th	301	310	291	2,879.7	7,007.3	15,365.3	9,027	70,505	189,508
99th	942	873	796	8,816.4	18,709.7	44,205.7	33,837	184,734	385,781
Quantile Ratios									
10th	0.029	0.032	0.033	0.013	0.011	0.020	0.000	0.033	0.033
20th	0.188	0.206	0.217	0.106	0.087	0.091	0.000	0.139	0.129
30th	0.290	0.365	0.367	0.237	0.202	0.211	0.036	0.219	0.195
40th	0.464	0.540	0.500	0.420	0.374	0.319	0.244	0.326	0.364
50th	0.681	0.762	0.683	0.548	0.537	0.469	0.414	0.618	0.531
60th	1.000	1.000	1.000	1.000	1.000	1.000	1.000	1.000	1.000
70th	1.652	1.841	1.800	1.774	1.539	1.491	1.526	1.596	1.474
80th	2.855	3.079	2.883	3.602	2.802	2.624	3.808	2.287	2.211
90th	4.362	4.921	4.850	5.844	5.553	5.082	24.732	4.764	4.500
99th	13.652	13.857	13.267	17.890	14.827	14.622	92.704	12.484	9.161

Quantile Ratios = Qth-Quantile/Median

Table 7 shows that the ranking patterns of geographic areas according to the number of sites and total amount of MIPs seems to be quite stable over time. This is easy to explain since the unit of observation is not scale free. That is, despite rapid rates of growth overall, the bigger areas are still bigger and the smaller areas still smaller. Hence, this is some evidence that the networking revolution did not alter the geographic distribution of computing stocks in any dramatic way.

Table 7 : Spearman Correlation Matrix Geographic Distribution Computer and Fiber Optic Infrastructures

		No. of Sites			All Systems Mips			Fiber Optic Cable		
		1986	1989	1992	1986	1989	1992	1986	1989	1992
No. of Sites	1986	1.00	1.00	1.00	0.97	0.97	0.97	0.72	0.86	0.85
	1989		1.00	1.00	0.97	0.97	1.00	0.73	0.87	0.86
	1992			1.00	0.97	0.97	0.97	0.73	0.86	0.86
All Systems Mips	1986				1.00	0.99	0.98	0.73	0.85	0.85
	1989					1.00	0.99	0.72	0.85	0.84
	1992						1.00	0.74	0.86	0.85
Fiber Optic Cable	1986							1.00	0.83	0.83
	1989								1.00	0.94
	1992									1.00

* All coefficients have a p-value of 0.0001

Fiber optics, however, do not show an identical pattern. The ranking pattern of geographic areas has changed between 1986 and 1989. Closer examination, not shown in the tables but revealed by the raw numbers in the appendix of GLS [1997], shows that the changes in the ranking pattern are due to changes in the deployment of fiber in the geographic areas that occupy the mid-size positions, and not in the areas that occupy the extreme positions of the ranking pattern. That is, areas that had a very small or a very large deployment of fiber optic in 1986 continue holding their position in the 1992 ranking pattern, but there was much growth among LECs with middle size fiber deployments. We conjecture that a rearrangement of investment patterns in the post-divestiture era altered in the ranking of regions in infrastructure.

The Gini coefficient and the Decile Distribution Ratio, shown in Table 8, reveal that the level of the two computer infrastructure variables show a slight reduction in inequality among regions.[34] The Decile Distribution Ratio corresponding to Number of Sites is almost twice the ratio for All Systems Mips, implying that the presence of users of large-scale computers has become more concentrated in regions where the presence had been low, but these users are basically small capacity users.

[34] The Gini coefficient of the geographic distribution of number of sites decreased from 0.678 in 1986 to 0.662 in 1992, while the Gini coefficient of the distribution of All Systems MIPs decreased from 0.718 in 1986 to 0.702 in 1992.

Table 8: Measures of Inequality of the Geographic Distribution Computer and Fiber Optic Infrastructures

	1986	1989	1992
GINI COEFFICIENTS			
Number of Sites	0.667	0.669	0.661
All Systems Mips	0.718	0.708	0.702
Fiber optic	0.856	0.671	0.651
DECIL DISTRIBUTION RATIOS			
Number of Sites	0.056	0.060	0.066
All Systems Mips	0.031	0.035	0.037
Fiber optic	0.003	0.049	0.054

The changes in the distribution of computing capacity at the regional level are surprising, considering the increasing skewness of computing capacity at large-scale computing sites. That is, the sites themselves are becoming more concentrated even though regions are not.[35,36] The Decile Distribution Ratio shows the same pattern. Further, there is a reduction in inequality of Fiber Optic cable. Although the major changes occurred from 1986 to 1989, the reduction of the inequality in the geographic distribution of fiber optic deployment continued to 1992.[37]

In summary, advanced IT grew tremendously over this time period. The degree of agglomeration declined along with that growth (though IT infrastructure did continue to agglomerate in urban areas at any point in time). Much of this change in the degree of agglomeration is due to growth in fiber at many localities, but some of it also appears to be due to the geographic spread of computing capital (despite the reduction in the number of sites). Since these are not scale-free observations, we are cautious about inferring much about changes to the access to advanced IT, which we analyze below. However, we note that the large decline of any areas with "no fiber," principally among smaller rural areas, suggests that we will find an increase in access to infrastructure at the lower end of the distribution.

2. Changes to the Distribution of Access to Information Infrastructure

We turn now to our analysis of changes to the access to information infrastructure, a more scale-free measure of the distribution. Table 9 compares the quantile ratios in 1992, 1989 and 1986. We find that for all indices almost all of the quantiles are getting closer to the

[35] A smaller fraction of the sites is holding a large percentage of total MIPS. The Gini coefficient of the site distribution of All Systems MIPs (across all sites), presented in Table 5, was 0.729, 0.733, and 0.750 in 1986, 1989, and 1992, respectively.

[36] Just 25 local exchange operating areas show a reduction of the Gini coefficient of the site distribution of All Systems Mips from 1986 to 1992, 34 from 1986 to 1989, and 33 from 1989 to 1992. In only 8 local exchange operating areas the Gini coefficient of the site distribution of Total Mips systematically decreased during the years 1986, 1989 and 1992. These places were United in Florida, Cincinnati Bell in Kentucky, United in Missouri, North western Bell in Nebraska, New Jersey Bell, United in Pennsylvania, United in Tennessee, and US WEST in Wyoming.

[37] The Gini coefficient decreased from 0.855 in 1986 to 0.67 and 0.65 in 1989 and 1992, respectively.

median in 1992. Much of this is due to greater diffusion of information technology in geographic areas where the access to this technology used to be low, but some is also due to declines in the relative lead of areas where infrastructure is high.

Table 9: Quantiles and Quantile Ratios Geographic Distribution Indices of Access to Computer and Fiber Optic Infrastructure

	Fiber Optic / Land			All Systems MIPS / POP			Number of Sites / POP			Net-Working Index		
	1986	1989	1992	1986	1989	1992	1986	1989	1992	1986	1989	1992
Quantiles												
10TH	0.000	0.044	0.253	0.032	0.084	0.137	0.016	0.017	0.014	0.000	0.006	0.017
20TH	0.000	0.184	0.660	0.094	0.248	0.525	0.024	0.026	0.022	0.000	0.043	0.416
30TH	0.001	0.340	1.117	0.138	0.375	0.887	0.031	0.035	0.030	0.000	0.123	0.679
40TH	0.009	0.455	1.404	0.229	0.491	1.253	0.040	0.041	0.039	0.001	0.214	1.336
50TH	0.014	0.643	1.775	0.320	0.749	1.661	0.048	0.047	0.044	0.003	0.406	2.678
60TH	0.024	0.854	2.409	0.423	0.949	2.065	0.055	0.053	0.050	0.009	0.578	4.142
70TH	0.044	1.401	3.477	0.505	1.163	2.415	0.062	0.060	0.055	0.023	1.146	7.784
80TH	0.085	2.749	6.194	0.624	1.440	3.036	0.071	0.067	0.059	0.049	2.801	15.173
90TH	0.250	5.367	11.186	0.733	1.771	3.605	0.079	0.078	0.069	0.109	7.485	32.702
99TH	2.161	14.679	47.038	2.036	8.145	14.049	0.593	0.588	0.464	1.172	59.077	260.784
Quantile Ratios												
10TH	0.000	0.016	0.041	0.051	0.058	0.045	0.223	0.252	0.236	0.000	0.002	0.001
20TH	0.000	0.067	0.106	0.151	0.172	0.173	0.344	0.381	0.371	0.000	0.015	0.027
30TH	0.009	0.124	0.180	0.221	0.261	0.292	0.437	0.513	0.513	0.001	0.044	0.045
40TH	0.104	0.165	0.227	0.367	0.341	0.413	0.559	0.607	0.657	0.023	0.076	0.088
50TH	0.160	0.234	0.287	0.513	0.520	0.547	0.679	0.702	0.741	0.056	0.145	0.176
60TH	0.275	0.311	0.389	0.678	0.659	0.680	0.773	0.781	0.851	0.189	0.206	0.273
70TH	0.513	0.510	0.561	0.810	0.808	0.796	0.869	0.893	0.932	0.468	0.409	0.513
80TH	1.000	1.000	1.000	1.000	1.000	1.000	1.000	1.000	1.000	1.000	1.000	1.000
90TH	2.932	1.953	1.806	1.175	1.230	1.187	1.117	1.156	1.174	2.232	2.673	2.155
99TH	25.298	5.340	7.594	3.265	5.656	4.628	8.347	8.742	7.867	24.066	21.094	17.187

*Washington D.C. is excluded.

In Table 10 we observe a high correlation among the index of access to computing technology and the index of power of computing capacity (0.92, 0.90, and 0.88 in 1986, 1989, and 1992, respectively). Similarly, the correlation over time of Number of Sites/ Population shows that for this access index the ranking of geographic areas stayed basically the same over the years. The small changes that have occurred in the ranking pattern of All Systems MIPs/Population are due to changes in the rank of those regions that are between the 20th and 80th percentile of the distribution. That is, the position of the smallest stays the same over time, as well as the ranking of the largest. Only the middle rankings have changed significantly.

Table 10: Spearman Correlation Matrix of Indices of Access to Computer and Fiber Optic Infrastructures

		Total Mips/ Pop			No. of Sites/ Pop			Fiber Optic/ Land			Net-working Index		
		1986	1989	1992	1986	1989	1992	1986	1989	1992	1986	1989	1992
Total Mips/ Pop	1986	1.00	0.97	0.92	0.92	0.91	0.86	0.48	0.50	0.56	0.60	0.77	0.80
	1989		1.00	0.94	0.92	0.90	0.87	0.45	0.49	0.54	0.56	0.77	0.80
	1992			1.00	0.87	0.88	0.88	0.44	0.43	0.47	0.54	0.71	0.78
No. Sites/ Pop	1986				1.00	0.98	0.93	0.43	0.50	0.55	0.54	0.73	0.77
	1989					1.00	0.96	0.44	0.50	0.55	0.55	0.73	0.77
	1992						1.00	0.39	0.45	0.49	0.49	0.67	0.73
Fiber / Land	1986							1.00	0.64	0.66	0.97	0.67	0.68
	1989								1.00	0.94	0.65	0.89	0.22
	1992									1.00	0.68	0.88	0.86
Net-working Index	1986										1.00	0.74	0.75
	1989											1.00	0.96
	1992												1.00

*All coefficients have a p-value of 0.0001.

Now we examine the Fiber/Land access index. The ranking in 1986 is different from the ranking pattern at the end of the period. Table 10 also shows that the places that have higher levels of Fiber/Land tend to be, in general, places with higher levels of Number of Sites/Population and All Systems MIPs/Population.[38] Thus, it is not surprising that the networking index shows that over time many regions contain a larger level of access to information infrastructure. Furthermore, the networking index indicates that the differences between regions are getting smaller.

Table 11 shows that most of the indices seem to have become more equally distributed across the period. As with our analysis above, the skewness in these data are somewhat deceptive. Going back to Table 2, we observe that most of the U.S. population lives within territories that are in the upper quartile. Most of the under-performing territories are also relatively unpopulated.

[38] In particular, the correlation among the Number of Sites/Population and Fiber/Land are 0.43, 0.52, and 0.48 in 1986, 1989, and 1992, respectively. The correlation among All Systems MIPs and Fiber/Land is around 0.48 in the three years.

Table 11: Measures of Inequality of the Geographic Distribution Indices of Access to Computer and Fiber Optic Infrastructures

	1986	1989	1992
GINI COEFFICIENTS			
No. Sites/Population	0.419	0.404	0.389
All Systems Mips/Population	0.459	0.494	0.464
Fiber/Land	0.851	0.680	0.657
Networking Index	0.868	0.826	0.796
Relative Information Infrastructure Index	0.033	0.036	0.035
DECIL DISTRIBUTION RATIOS			
No. Sites/Population	0.354	0.388	0.408
All Systems Mips/Population	0.222	0.200	0.237
Fiber/Land	0.005	0.059	0.084
Networking Index	0.001	0.010	0.144
Relative Information Infrastructure Index	1.844	1.829	1.831

In summary, there is less agglomeration over time in access to advanced information technology capital. This change in access is due to changes in the dispersion of fiber optics and computing. The changes to fiber are a post-divestiture phenomenon, so these results raise questions about the determinants of LEC investment in the post-divestiture era. The changes to computing occurred in spite of a gradual decline in the number of sites using large scale computing. This raises issues about the influence of the networking revolution on the regional dispersion of advanced IT. Finally, the high correlation in the location of these two components of infrastructure motivates a further examination of the co-determinants of these complementary inputs.

VI. Conclusion

In this study, we have examined the geographic distribution of advanced information infrastructure in computing and telecommunication between 1986 and 1992, a period of rapid growth. In all our estimates, we interpret the level of capital stock as a proxy for a broad variety of activities associated with building public and private digital IT networks in a region. Our findings may be summarized as follows:

- Advanced infrastructure grew at a tremendous rate in the late 1980s and early 1990s across all regions. Virtually every region experienced growth in advanced infrastructure and increasing access to it.
- Advanced IT grows rapidly in dense urban areas. Yet, density alone does not explain the location patterns of advanced IT.
- Areas with larger white-collar sectors also have more advanced information infrastructure. Information intensive industries also tend to be more geographically

dispersed than less information intensive industries, potentially contributing to the spread of advanced IT.

- Higher income areas and areas experiencing recent growth in urban population do not have more agglomeration of advanced IT.
- The US experienced changes in the distribution of infrastructure capital during a period of rapid growth. The large changes did not alter the relative rankings of the "richest" or the "poorest" regions, but did alter the relative disparity between them. It also altered many of the rankings of regions in the middle.
- During this period, increasing equality of access to advanced information infrastructure was not due to extreme changes in the ranking of regions. It was due primarily to the increase in the relative standing of regions on the bottom and relative declines in the regions at the top.

These conclusions need to be stated carefully. We cannot argue on the basis of available evidence that the distribution of advanced IT capital matches some policy-relevant notion of an appropriate distribution. Indeed, no such distribution can be precisely defined for concrete data since there are inherent ambiguities in measuring the distribution of infrastructure and access to it. However, we can conclude that the period prior to the commercialization of the Internet was one of rapid growth in advanced IT in the U.S. Furthermore, we can conclude that this period did not lead to an increased agglomeration of advanced IT capital. To the contrary, the evidence suggests increased dispersion. That is, all regions, rural and urban, whether initially ahead or behind, became more alike over time. By virtually any ideal, this movement is a good thing.

Other economic changes follow the growth in advanced information infrastructure, its standardization and its decline in price. Independent service providers, broadband access, forms of electronic commerce, and support services for digital communications and networking, for example, all work better with advanced telecommunications networks. Many business services tied to local labor markets for technical help will also be aided. On-line-transaction-processing of large data bases, such as credit checking, automated inventory restocking and reservation systems, all require advanced large-scale computing facilities to run well. Our analysis gives a positive conclusion to concerns about regional disparities in the backbone to advanced IT. Our findings suggest that only a small number of areas, in particular small and less densely populated regions, may not have direct access to advanced IT capital. Furthermore, most areas have improved over time, easing the diffusion of nationwide services and reducing the degree of disparities across regions.

Our results call for a more careful modulation of the debate regarding the redefinition of universal service in the new information infrastructure. The agglomeration of advanced IT followed a somewhat predictable economic logic -- locating in areas in relation to a few features of those areas, such as the presence of IT intensive industries, high income and density. The relative lead of the earliest investors did not seem to last for very long. It also seems that in this instance investment has decreased the concentration of advanced IT over the period. Further research of these patterns needs to understand the micro-economic determinants of investment in post-divestiture telecommunications and networking. Further work may also need to carefully focus on specific geographic areas of low access or specific populations whose experiences cannot be reflected in the data analyzed in this study.

REFERENCES

Bresnahan, T. and S. Greenstein [1996], "The Competitive Crash in Large-scale Commercial Computing" in (Eds) Ralph Landau, Timothy Taylor, and Gavin Wright, *The Mosaic of Economic Growth.* Stanford, CA. Stanford University Press.

Bresnahan, Timothy, and Shane Greenstein [1997]. Technical Progress in Computing and in the Uses of Computers. *Brookings Papers on Economic Activity, Microeconomics,* 1-78.

Bresnahan, T. and S. Greenstein [1999], "Technological Competition and the Structure of the Computing Industry," *Journal of Industrial Economics.*

Bresnahan, Timothy and Manuel Trajtenberg [1995] "General Purpose Technologies: 'Engines of Growth'?," *Journal of Econometrics*, Special Issue, January, v 65, n 1, pp. 83-108.

Brody, Herb [1988], "The Rewiring of America" *High Technology Business,* February. pp. 34-38.

Brynnolfsson, E. and L. Hitt [1995], "Information Technology as a Factor of Production: The Role of Differences among Firms." *Economics of Innovation and New Technologies*, Vol. 3, pp. 183-199.

Cairncross, Frances., [1997]. *The Death of Distance. Cambridge, MA:* Harvard University Press.

Capello, R. [1994], *Spatial Economic Analysis of Telecommunications Network Externalities*, England: Avebury.

Cowell, F., [1995] *Measuring Inequality.* 2nd Ed. Prentice Hall.

T. Downes and S. Greenstein [2002], "Universal Access and Local Internet Markets in the US," *Research Policy,* 31, pp1035-1052.

Forman, Chris [2002], *The Corporate Digital Divide: Determinants of Internet Adoption.* Mimeo. Carnegie Mellon University.

Forman, C., A. Goldfarb, S. Greenstein [2002], "Digital Dispersion: An Industrial and Geographic Census of Commercial Internet Use," *NBER Working Paper 9287.*

Friedman, A. with D. Cornford [1989], *Computer Systems Development: History, Organization, and Implementation,* Chichester, England; New York: Wiley.

Greenstein, S. [2000], "Commercialization of the Internet: The Interaction of Public Policy and Private Choices or Why Introducing the Market Worked so Well." In (eds) A. Jaffe, J. Lerner, and S. Stern, *Innovation, Policy and the Economy.* MIT Press.

Greenstein, S. and M. Lizardo [1997], "Determinants of the Regional Distribution of Information Technology Infrastructure in the United States," in (eds) D. Orr, and T. Wilson, *The Electronic Village, Policy Issues of the Information Economy, C.D. Institute, Ottowa, Ontario, Canada.*

Greenstein, S., M. Lizardo, and P. Spiller [1997], "The Evolution of Advanced Large Scale Information Infrastructure in the United States," National Bureau of Economic Research, *Working Paper # 5929.*

Greenstein, S., S. McMaster, and P. Spiller [1995], "The Effect of Incentive Regulation on Local Exchange Companies' Deployment of Digital Infrastructure," *Journal of Economics, Management, and Strategy.*

Greenstein, S. and P. Spiller [1996], "Estimating the Welfare Effects of Digital Infrastructure." National Bureau of Economic Research, *Working Paper # 5770,* September.

Hart, B., A. Nave, A. Raskob Jr., and J. Thomason [1982], *Telephone Areas Serviced by Bell and Independent Companies in the United States.* U.S. Dept. of Commerce. NTIA Report 82-97.

Hepworth, M. [1995], *Geography in the Information Economy*, The Guilford Press, New York.

Helpman, Elahanan [1998]. *General Purpose Technologies and Economic Growth.* Cambridge, MA: The MIT Press.

Information Infrastructure Task Force [1993], *The National Information Infrastructure: Agenda for Action.*

Information Infrastructure Task Force [1994], *National Information Infrastructure: Progress Report September 1993-1994.*

Kahin B. [1991], *Building Information Infrastructure: Issues in the Development of National Research and Education Network.* McGraw-Hill Primis.

Kahin, B. and J. Keller [1995], *Public Access to the Internet*, The MIT Press: Cambridge, MA.

Koenker, R. and G. Bassett Jr.[1978], "Regression Quantiles," *Econometrica,* Vol. 46, No. 1, pp. 33-50.

Kolko, Jed [2002]. "Silicon Mountains, Silicon Molehills, Geographic Concentration and Convergence of Internet Industries in the US," *Economics of Information and Policy.*

Kotkin, Joel [2000], *The New Geography: How the Digital Revolution is Reshaping the American Landscape*, New York: Random House.

Krugman, P.[1991], *Geography and Trade.* Leuven University Press, The MIT Press.

Lavoie, Francis, J [1988], "Digital Technology Fuels the Phone Market" *Modern Office Technology,* 33, 5, May, pp. 14-18.

Markusen, Ann, Peter Hall, and Amy Glasmeier [1986], *High Tech America: The What, How, Where and Why of the Sunrise Industries*, Boston, MA, Allen & Unwin.

McMaster, S. [1995], *Telecommunications in a new Era of Competition: Fiber Optics, Regulatory Barriers, and Competition.* Unpublished Dissertation, University of Illinois.

Minoli, D. [1991], *Telecommunications Technology Handbook*, Boston: Artech House, 1991.

Moss, Mitchell L., and Anthony M. Townsend [1997], Tracking the net: using domain names to measure the growth of the Internet in US cities. *Journal of Urban Technology* 4(3): 47-60.

Munell, A. H. [1992], "Infrastructure Investment and Economic Growth," *Journal of Economic Perspectives*, 6(4), Fall, pp. 189-198.

Mueller, M. [1993], "Telecommunications as Infrastructure: A Skeptical View," *Journal of Communications,* 43(2), Spring, pp. 147-159.

Mueller, M. [1997], *Universal Service: Competition, Interconnection, and Monopoly in the Making of the American Telephone System*, MIT Press, Cambridge, MA.

National Academy of Engineering [1995], *Revolution in the U.S. Information Infrastructure.* Washington, D.C., National Academy Press.

National Information Infrastructure Advisory Council [1995], *Common Ground: Fundamental Principles for the National Information Infrastructure.*

National Research Council [1996], *The Unpredictable Certainty: Information Infrastructure Through 2000.* Washington D.C.: National Academy Press.

National Telecommunications and Information Administration [1995], *Connecting the Nation: Classrooms, Libraries, and Health Care Organizations in the Information Age.*

Nissan, E. [1994], "A Composite Index for Statistical Inference for Ranking Metropolitan Areas' *Growth and Change*," Vol. 25, Fall, pp.411-426.

Radford, Bruce, W. [1989], "Fiber-optic Networks Nibble Away at the Bell Local Monopoly," *Public Utilities Fortnightly*, 124, 13, pp. 4. December 21.

Reese, Frank D [1988], "New Services -- What's in It for Us?" *Telephone Engineer & Management*, 92, 10, May 15, pp. 69-76

Roller H., and L. Waverman [2001], "Telecommunications Infrastructure and Economic Development: A Simultaneous Approach," *American Economic Review,* V91 (4), p909-923.

Saxenian, Annalee [1994], *Regional Advantage: Culture and Competition in Silicon Valley and Route 128,* Harvard University Press, Cambridge, MA.

Solomon, R. [1995], "Telecommunications Technology for the Twenty-First Century" in (Ed)

William Drake, *The New Information Infrastructure: Strategies for U.S. Policy.* The Twentieth Century Fund Press. New York.

Steinmueller, W. Edward [1996], "Technological Infrastructure in Information Technology Industries," in (Eds.) Morris Teubal, Dominique Foray, Moshe Justman and Ehud Zuscovitch, *Technological Infrastructure Policy: An International Perspective*, Kluwer Adademic Publishers, London, UK.

Teske, P. [1995], *American Regulatory Federalism & Telecommunications Infrastructure.* Hillsdale, NJ: Lawrence Erlbaum Associates.

Teubal, Morris, Dominique Foray, Moshe Justman and Ehud Zuscovitch [1996], *Technological Infrastructure Policy: An International Perspective*, Kluwer Adademic Publishers, London, UK.

U.S. Advisory Council on the National Information Infrastructure [1996], *A Nation of Opportunity: Realizing the Promise of the Information Superhighway.*

Zooks, Matthew [2000b], "The Web of Production: the Economic Geography of Commercial Internet Content Production in the United States," *Environment and Planning*, 32: 411-26.

Chapter 2

THE EVOLUTION OF TELECOMMUNICATIONS SWITCHING IN THE CENTRAL OFFICE*

Allan L. Shampine
Lexecon Inc.

I. INTRODUCTION

As telecommunications' impact on economies has grown, a variety of studies have tried to quantify the relationship between telecommunications infrastructure and economic growth (see Roller & Waverman 2001, Lichtenberg 1995, Roller & Waverman 1996, Norton 1992, and Hardy 1980). This literature is subject to the usual problems in identifying causality, but the studies have invariably found a strong correlation between telecommunications infrastructure investments and economic growth. Various public entities have recognized this correlation and instituted programs to promote the deployment of telecommunications technologies. For example, the World Bank's Information Development Programme disseminates policy advice on information and communication technologies for development and best practices. The International Telecommunications Union has established telecommunications training centers worldwide. The United Nations' Economic and Social Council created the ICT Task Force and Trust Fund to assist in deploying information and communication technologies. Also, the G8 states created the Digital Opportunities Task Force to match donors with projects promoting telecommunications technologies for development. (Campbell 2001)

All of these efforts require an understanding of how and why telecommunications technologies are adopted. If a policy maker wishes to speed up adoption of a technology, it must identify the bottlenecks that are constraining adoption. Furthermore, information on how firms adopt new technologies helps policy makers determine whether intervention is

* An earlier version of this paper appears as "Handicapping Countries in the Race to Digital Switching" in *Progress in Economics*.

I would like to thank William Lynk and David Gross for their helpful comments. The views expressed herein are those of the author and not necessarily those of Lexecon.

worthwhile. If a country has efficient capital markets, then the factors constraining adoption should provide insight into both whether and how to intervene.

This paper examines the evolution and diffusion of one of the basic building blocks of a telecommunications network – the central office switch. Section II discusses the history of central office switching and places the diffusion process in context by discussing the relative merits of different generations of switches. Section III examines the nature of the adoption decision with respect to switches and discusses the implied determinants of adoption. Section IV describes two data sets used to examine switch adoption: historical data for 19 developed countries from 1975 to 1998, and more detailed data from the United states from 1983 through 1996. Section V presents empirical results on the determinants of adoption. Section VI summarizes the findings and outlines several areas for future research.

II. History of Telecommunications Switching in the Central Office

This section presents a brief history of telecommunications switching. A telecommunications switch is a device that routes a signal between two points (such as two telephones). A central office switch is one that is located at the edge of a company's network and aggregates traffic from end users.[1] Advanced features are typically implemented at the central office level. These switches are also known as Class 5 switches following the old AT&T switching hierarchy. Higher level switches called tandem or toll switches aggregate traffic from central office switches. In the AT&T switching hierarchy, a switch which aggregated toll traffic from Class 5 switches was called a Class 4 switch. Class 5 switches are much more numerous than Class 4 switches – as of 2001 there were over 20,000 central office switches in operation in the United States as compared to roughly 750 Class 4 switches. (Theodosopoulos 2001) Central office switches are not only numerous, they are expensive. At $5 to $10 million each, decisions concerning switches are taken very seriously by companies. This paper focuses exclusively on the evolution and deployment of central office switches.

Central office switches have evolved through several stages – electromechanical, electronic circuit switched (analog then digital), and finally electronic packet switched. Originally, phone systems used human operated switches – the classic switchboard operator -- where an operator would ask for whom the call was intended and complete the circuit by hand. In most industrialized countries, however, these switchboards were replaced with electromechanical switches.

A. Electromechanical Switches

Electromechanical switches are automated versions of the human operated switchboard where circuits are physically closed without human intervention. Electromechanical relays physically move in response to the digits dialed by a caller. Three classes of these switches, varying in sophistication, were available by the late 1930s. These switches are inferior in

[1] A central office is simply a location where end-user lines are aggregated at a switch.

virtually every dimension to electronic switches. Nonetheless, they continue to exist in small pockets in telecommunications networks around the world although the number of lines served by them is extremely small. Typically, an electromechanical switch might remain when it serves a handful of customers with low demand for advanced services. Nonetheless, the operating savings available from electronic switches are substantial, and the few remaining electromechanical switches are being phased out.

Although the deployment of electromechanical switches is important historically, there are few reliable data on the subject. Most countries did not begin tracking switch and line deployment until after the development of electronic switches.

Electronic switches added 'intelligence' to electromechanical switches. The electronic switch has a computer 'brain' that allows it to perform operations and services that simple electromechanical relays cannot. The first generation of electronic switch created circuits that maintained the analog nature of the phone signal.[2]

B. Electronic Circuit Switches – the Analog Switch

The first true analog switch, the 1-ESS, was installed in 1965. This commercial central office switch had stored program control (software) but still depended on space-division switching (meaning that circuits, or phone connections, consist of physically discrete paths). One important difference from the older electromechanical switches was the addition of 'intelligence' to the switch – a portion of the switch that could perform more elaborate functions than the simple electromechanical relays of the previous generation. These early electronic switches are collectively known as analog switches because of their reliance on circuits that preserve the analog nature of the signal.

Electronic analog switches were more reliable, smaller and more efficient than electromechanical switches. The analog switches could service many more lines and could also provide services that were impossible to provision on an electromechanical switch. Analog switches quickly became the mainstay of the telecommunications industry and were widely deployed throughout the world. However, as sophisticated as these switches were in comparison to electromechanical switches, they still relied on analog circuits. Digital switches, by contrast, convert the analog signal into bits (0s and 1s) that can then be mixed together with other signals to make more efficient use of available circuits. The method of mixing used by digital switches is called time-division switching (also known as time division multiplexing or TDM) which allows multiple signals to share a single physical path.[3] The digital switches still use circuits, however – a call is allocated a certain amount of bandwidth that is guaranteed for the duration.

Upstream from the central office, high capacity backbones and frame relay or ATM "clouds" were created or converted to digital traffic, and data services that were transmitted digitally also began to grow in popularity. The development of digital services such as T-1

[2] Virtually all non-wireless telephones today generate an analog signal – a smooth and continuous electromagnetic signal. An analog circuit preserves this signal through its entire length, as opposed to a digital circuit, which converts the information in the analog signal into 0s and 1s (bits).

[3] With TDM, each conversation is sampled and the samples are sent in sequence. The switch assigns each call a set of time slots, then picks the data out of that slot and reassembles the call. The time division in the name refers to the fact that the data for each call are tracked through their assigned time slot (such as one call's data arriving at even numbered seconds and another call's data arriving at odd numbered seconds).

lines required the deployment of interfacing hardware and software that could convert signals from analog to digital and back again as the analog central office switches received and handed off traffic. Given these developments, it was inevitable that digital switches would move from the backbones into the central offices as well.

C. Electronic Circuit Switches – the Digital Switch

Commercial digital switching began in 1970 with the installation of an E10 electronic switching system in France. Bell Laboratories announced in 1972 that it would develop the No. 4 ESS switching system to be aimed at very large capacity offices. It became "the first commercial stored program controlled digital toll switch to be placed in service in the world in January 1976" (Joel, 1979). Digital switches were popular in developed countries with a high demand for telecommunications because of their more efficient use of the existing physical plant – a single line could carry many signals simultaneously. Also, analog switches required expensive conversion hardware when connected to digital transmission systems such as T-1 lines, which were becoming fairly widespread by the mid 1970s in countries such as the U.S. Digital switches were smaller than analog switches and required less power. They also did not require any conversion hardware for connecting to other digital systems.

Flamm (1989) estimated that replacing an electromechanical system with a digital stored program control system reduced lifetime maintenance costs by 40 to 60 percent of the switch's initial cost, or annual savings of 6 to 7 percent of the switch's cost, while "[t]he changeover from an analog switching matrix to a digital system in 1976 resulted in [lifetime] savings of 10 to 20 percent of initial cost," or annual savings equal to roughly 2 percent of the switch's cost.

It may sound like digital switches are uniformly better than their predecessors. This is not, in fact, the case. Flamm's estimates, for example, were developed with respect to the U.S. economy. Digital switches are not necessarily superior to analog switches for all countries or circumstances. A digital switch can serve more lines than an analog switch and can transmit multiple signals on a single line, but an analog switch may be more cost effective in areas of low demand (either because of low population density or low demand for telecommunications services), particularly if there is relatively little need to interface with other digital transmission facilities such as T-1s. Similarly, if a population has little demand for advanced features such as caller ID, then the ability of a digital switch to offer these services is largely irrelevant.

Adoption of digital technology is not necessarily an either/or proposition. Older analog systems such as the 1-ESS may be upgraded or used in conjunction with digital remote terminals. Furthermore, the switches themselves are not static entities. Over time, software has represented a greater proportion of the cost of a switch and existing switches may have their software upgraded repeatedly. Above and beyond the cost savings offered by digital switches, digital technology allows new services such as caller ID, voice mail and virtual call centers. Many of these services, made possible by the new generation of switches, are highly profitable in developed countries but almost unknown in other countries. Moreover, their profitability is only known ex post. Companies and countries trying to choose their upgrade path have to predict the demand for the new services as well as simply calculating cost savings.

D. Electronic Packet Switches

Digital switches increased efficiency over analog switches by allowing time division multiplexing (TDM). However, even though multiple calls could share a circuit, digital switches still required dedicated circuits, and each call required its own dedicated bandwidth. Packet switching, by contrast, does not require a dedicated circuit.

In packet switching, all traffic, whether voice calls or data, is digitized and then chopped up into pieces (packets). The switches or routers then allocate the pieces to make the most efficient use of the physical routes available. Historically, the difficulty of using packet switching for voice services is that the method provided efficient use of physical facilities at the expense of guaranteed timely delivery. A dedicated circuit guarantees that its traffic will arrive in a timely fashion, but often wastes bandwidth since the circuit must be maintained even when people are not speaking. Packet switching, on the other hand, will fill the available bandwidth efficiently, but historically the voice packets did not receive any preference over the data packets, and tiny delays that are negligible for data can be perceived by humans as lag in their phone calls. Recent innovations in packet switching technologies have addressed this problem by allowing packets to be given different priorities. Voice packets can thus be given priority over data, reducing or eliminating latency.

Packet switching, while ubiquitous in backbones and data networks, remains rare in central offices. While it is generally agreed that all circuit switches will eventually be replaced with packet switches of one form or another, it is not clear when this will occur, particularly given the depressed circumstances of the industry today.

III. The Adoption Decision

When a firm has the choice of installing a new switch it asks whether the net present value of installing that switch is positive. The overall level of adoption is simply the aggregate of these individual decisions. In this section I will discuss the factors that affect these adoption decisions.

Over the past few decades this choice has largely been limited to analog and digital switches. In the last few years a handful of central office packet switching solutions have been deployed, but they do not appear in the data studied here.

Regulation

Regulation is not a determinant of adoption in and of itself, but it can affect many aspects of the adoption decision. Regulation can affect a firm's objective function (whether it is maximizing profits or not), its costs, its revenues, and its investment options. The effects of regulation on telecommunications diffusion have been found to be important by Greenstein, McMaster & Spiller (1995), Berg & Hamilton (2001), and Gutierrez & Berg (2001). These studies also illustrate the difficulties of quantifying distinctions between regulatory regimes. That is, regulatory regimes are so varied and idiosyncratic that it is extremely difficult to quantify distinctions between them in any way except as a fixed effect. Such comparisons are particularly difficult between different countries.

The effects of regulation on individual aspects of the adoption decision are discussed in the appropriate sections. However, there is one overriding element to be mentioned here – the ownership of the telecommunications infrastructure. Many countries exercise direct control over their telecommunications networks through regulated monopolies. Other countries have privatized their systems, although every country has retained at least some degree of regulation. In each case the telecommunications company cares about the direct costs of deployment. However, government controlled entities may not be maximizing profits. If they are interested in maximizing consumer welfare, then they may be better able to internalize network externalities than private companies, but they may also have rather different objective functions (such as maximizing welfare for a particular voter bloc, for example). Also, it is unlikely that a regulated monopoly will set price at the competitive level – inadequate regulation might lead to a higher price, as would the use of telecommunications to subsidize other programs, or telecommunications might itself be subsidized, leading to a price below the competitive level. The effects of price on adoption are discussed later in this section.

Costs

The relative prices of the alternative technologies are determinants of adoption. On a cost per line basis, the relative cost effectiveness of analog and digital switches depends on the level of demand. That is, installing an analog switch is cheaper than installing a digital switch, but the digital switch can serve more customers and offer more services. Regulation and government action may also impact the costs that a firm faces. For example, some countries have implemented unbundling regulations that require incumbents to make their facilities available to competitors at regulated rates. The expected cost of such facilities is thus raised by the degree to which the regulated rates do not adequately capture the costs and risks born by the deployer. (Hausman & Sidak 1999)

Regulations may also affect costs through subsidies, accounting rules, or rate regulations. For example, in some countries certain types of deployment receive favorable tax status (such as broadband deployment in South Korea). In the United States the level of depreciation for existing technology was crucial under rate of return regulation. As Paul Travis explained in *Telephony* in 1989, the baby Bells "have billions and billions of dollars invested in the dependable and feature-rich 1A electronic analog switches, which work just fine and are not yet fully depreciated." Greenstein, McMaster & Spiller (1995) explained that such firms often have low levels of depreciation, mandated by regulators, which reduce the firms' incentives to invest in new technologies.

New entrants face a similar decision – whether to enter today using circuit switching or to wait and enter in the near future using packet switching. This is the situation that the cable television industry has been faced with in the United States in the last several years. Some companies, notably AT&T Broadband and Cox, have chosen to enter using circuit switching despite expectations that voice over packet switching capital costs would be 10 to 20 percent less than the circuit switching costs. (AT&T Broadband 2001) AT&T Broadband in particular had significant expertise with circuit switching that likely reduced the differential compared to other cable companies which have chosen to wait for packet switching to become commercially available. In essence, AT&T Broadband decided that the cost savings from waiting were outweighed by the revenues it would forego. Other companies have made different decisions.

The physical structure of the telecommunications network may also affect costs. For example, countries whose populations largely reside within dense population centers will also find digital switches relatively more cost effective than countries whose populations are more geographically diffuse. The size of a country and the distribution of its population are thus determinants of adoption as well.

Demand for Telecommunications Services

In general, firms deploy more advanced switches in areas with higher demand. For switches the relevant measure of demand is the number of lines required in a given geographic area. In part this is because digital switches are more cost effective in high density areas, and in part this is due to the demand for the more advanced services available from those switches. Dense metropolitan areas tend to contain business centers with higher demand for digital services as well. Demand is not directly observable but a variety of factors are correlated with the demand for telecommunications services. Perhaps the most direct measure is a waiting list for new services. Some countries, particularly those who do not yet have near universal POTS available, maintain waiting lists for phone lines. The ratio of the waiting list to the number of people who do not yet have service provides an estimate of demand. However, the mere existence of the waiting list demonstrates that the country has chosen not to fully deploy its network. Since that choice was endogenous, the country may delay deploying digital technology as well. On the other hand, the country may "leapfrog" if the incremental gains are larger for it than for countries that had invested in more modern or more extensive infrastructure.

Thus the distribution of population in a firm's area should affect the rate of adoption. However, the average population density is the wrong measure to capture this effect, particularly for large geographic areas. The level of urbanization or some form of GINI coefficient are more appropriate. In addition, the rate of growth in metropolitan areas should also affect the demand for new services and the expectations for costs over coming years. Rapidly growing urban areas would be expected to have more digital switches installed than decaying cities.

At a country level, demand is likely to be correlated with the gross domestic product. As Greenstein & Spiller (1995) have demonstrated, certain types of GDP tend to be associated with higher levels of demand, but such detailed data are not always available.

Finally, demand is affected by price. Service revenues per line provide a measure of price, but this price is not always set in a competitive market. Higher revenues per line may be indicative of greater demand, all else equal, but it is also possible that those higher revenues are due to supply being constrained below the competitive level. In that case, the effect on adoption is ambiguous – demand will be lower, but revenues per line would be higher. The lower demand may impact adoption through reduced network externalities. Also, since digital technologies need a high density of lines to be cost effective, it is plausible that countries constraining supply would adopt more slowly than countries pricing closer to the competitive level.

Network Effects

Network effects affect adoption through their impact on the quality of services being offered. The value of most telecommunications services is largely a function of the number of subscribers to the service or the level of deployment of the service – this is known as a network effect, or network externality. The addition of one more subscriber or digital switch makes the service more valuable to other subscribers. Similarly, the number of existing subscribers determines the value of the service to new subscribers. Thus the installed base of the firm (and the country) is a determinant of adoption because of its impact on demand through network externalities.

The network effects from the installed base include both technology specific and non-specific externalities. That is, a new switch provides some benefits to customers of all other switches, regardless of type. In addition, that switch may provide more benefits to customers using other switches of the same type. Adding a new digital switch will make basic services such as plain old telephone service (POTS) more valuable to all telecommunications customers simply because they can now reach more people. In addition, customers using other digital switches benefit because the digital specific services can also reach more people. For example, caller ID only functions if a particular call passes through switches that support that capability.[4] If it passes through an analog switch that does not support caller ID, the relevant piece of data will be stripped from the signal.[5] Similarly, it is easier to integrate digital services such as T-1 lines with digital switches than with analog switches.

The impact of these externalities on adoption, however, is unclear. Rosenkranz (1997) finds that low-quality firms (using a first generation technology) have a greater incentive to adopt a third generation technology in the face of significant adoption costs than would a firm using a second generation technology. However, Farrell and Saloner (1986) point out that a large installed base with network externalities can either decrease adoption because early adopters bear a larger proportion of the transition costs, or increase adoption as users of the older technology change to prevent being "stranded." Witt (1997) takes exception to the idea that "lock in" is significant, but concludes that a critical mass of adopters is required for a new technology to be successful. Shampine (2001) finds that in the U.S. telecommunications companies with large installed bases of analog switches adopted digital technology more slowly than companies with a preponderance of electromechanical switches.

The ambiguity of network effects is largely due to the endogenous nature of the installed base. At some point in the past the firm had to decide what type of switches to install or replace. Some firms chose to install or keep electromechanical or analog switches. The most likely reasons for such a choice include low population density and/or a perceived lack of demand for new features and services.[6] Therefore, as a matter of theory, a firm with a large

---

[4] That is, all of the switches must be using Signaling System 7 (SS7), a signaling protocol optimized for digital switches.

[5] This particular example is an over-simplification of both caller ID and the capabilities of analog switches. However, it is intended to make the point that many services are dependent upon the installed hardware and software, and that legacy hardware and software may impede their functionality.

[6] Industry participants sometimes try to explain slow adoption as a function of capital constraints. However, given efficient capital markets this explanation seems unlikely. If the perceived present value of adding new switches were positive, the capital should be available. In non-market economies or economies with significant corruption, then misallocation of resources becomes a more serious issue. Trade barriers can have a similar effect.

installed telecommunications base but little digital technology may or may not adopt digital technology more quickly than a new entrant or country with little in the way of infrastructure. The choice of electromechanical technology over analog or analog over digital demonstrates that a firm anticipates relatively lower demand for telecommunications services in general and advanced services in particular. Similarly, a firm which adopts analog switches over electromechanical, or digital over analog, or that waits for packet switching has shown that it expects a relatively high pay-off from the newer technology. Thus it is not entirely clear whether the expected gain from installing a digital switch is greater to a firm with no switch or an electromechanical switch than to a firm with an analog switch.

Firm Size

Firm size has been found to be positively correlated with the rate of adoption in virtually every industry studied. For example, in an agricultural context, Feder, Just & Zilberman (1985) concluded that the high yielding crop varieties of the Green Revolution were adopted faster by larger farms.[7] Rose and Joskow (1990) found that larger electric utilities adopt new technologies faster, and Dougherty, Germain and Druge (1995) found a similar result for the logistics management industry. Saloner and Shepard's 1995 study of ATM adoption by banks found that the number of branches was positively correlated with the rate of adoption. They attributed this correlation to the presence of network effects. As discussed, the value of new switching rises when all switches upgrade. However, the most that any one telephone firm can do is upgrade its own switches, while the benefits from their upgrade are realized by other interconnecting firms. Hence, private incentives do not internalize all the benefits from upgrading one switch, and the size of a firm's network determines how much of the network effects it can internalize. Majumdar and Venkateraman (1998) examined the relation of switch adoption and network externalities and attributed a positive correlation between firm size and the rate of adoption of electronic switches (both analog and digital) in the telecommunications industry to network effects. See also Rogers (1995) for a discussion of the relative adoption speeds of interactive innovations.

Installing a new generation of switches also requires the development of procedures and human capital in maintaining and operating those switches. To the extent that these costs are independent of firm size, larger firms could amortize the costs over a larger base of switches. Since telecommunications firms generally make significant efforts to establish corporate-wide network maintenance procedures, it is certainly plausible that, on a per switch basis, larger firms would spend less on updating their internal procedures and human capital.

Fixed Effects

Finally, there are a variety of firm specific effects that are difficult to capture independently. These include regulation, corporate focus and strategic plans, and complementary assets or plans. Regulation can affect adoption in many ways. For instance, it can affect the prices (and therefore the profits) that can be made from new and existing services. It can affect how and whether capital expenditures are made. It can affect the

[7] Although Ruttan has argued that smaller farms caught up with larger farms within a few years.

deployment and marketing of new services. Corporate focus and strategic plans reflect whether the firm is putting its money behind the services available through digital switches or whether it believes that it can obtain a greater return through some other avenue. This is tied in to complementary assets or plans. For example, a firm might be trying to establish a wireless presence and find that it is easier time to integrate its wireless operations with its analog switches. In that case the firm might delay installing digital switches in order to pursue the more immediate opportunity perceived in wireless.

IV. DATA

Two data sets are used to examine the determinants of switch adoption. A set of international data cover 21 OECD countries from 1975 to 1998. The countries included are Australia, Austria, Belgium, Canada, Denmark, Finland, France, Germany, Greece, Ireland, Italy, Japan, the Netherlands, New Zealand, Norway, Portugal, Spain, Sweden, Turkey, the United Kingdom and the United States. These data are drawn from the Summers and Heston 5.6 and 6.0 beta tables and the International Telecommunications Union (ITU) STARS database. Table 1 provides the names, definitions and sources of the variables used. Different countries begin reporting data in different years, and countries may begin reporting different series in different years as well. In total, the data set provide 381 usable observations.

Most of the variables are straightforward. However, while the ITU tracks telecommunications investment and Summers & Heston track total capital stock, neither examines telecommunications capital. For this analysis, a telecommunications capital series is constructed following Roller & Waverman (2001), using the perpetual inventory method with a 10 percent depreciation rate and initial values set so that the telecommunications stock is the same fraction of total capital stock as telecommunications investment is to total investment for the initial year.

Total labor force is used as a proxy for overall business demand. Prices for analog and digital systems are included where they are available historically, such as in the U.S., but in general transaction price data are lacking. Similarly, the number of digital, analog and electromechanical lines is used to model the installed base where the data are available. Where these data are not available, I use the stock of telecommunications capital as a proxy for previous investments. The higher the capital stock, the larger and more sophisticated the installed base is likely to be. I also use total government expenditures as a proxy for the level of government intervention. Finally, country specific fixed effects are included to capture the effects of regulation and other elements for which data are not available.

The international data are compared with results presented in Shampine (2001) which used relatively detailed data for multiple telecommunications firms in the United States. These data are compiled from various editions of the U.S. Central Office Equipment Market report (graciously provided by DataQuest) and from noted public sources. The data consist of holding company level information for fifteen companies (after mergers and acquisitions) on digital, analog and electromechanical lines, systems and prices from 1983 to 1996. The final companies included are Alltel, Ameritech, Bell Atlantic, BellSouth, Cincinnati Bell, Frontier, GTE/Contel, Lincoln (now Aliant), NYNEX, Other Independents, Pacific Telesis, SNET, SBC, Sprint/United and U S WEST.

In the U.S. data, firm size is modeled through the number of total lines owned by the company in the previous year. The installed base is captured by the percentage of lines that were electromechanical in the previous year. The distribution of population is captured through the percent of population in a firm's service area that lives in metropolitan areas. Finally, the change in the percentage of metropolitan population is intended to capture whether urban areas are growing or decaying. Decaying areas would display less potential for revenue growth.

Table 1.
Variable Definitions

International Data Set Variables

Log Govt. Expenditures

Log of total government expenditures (billions of 1996 dollars), by country. Source: Summers & Heston.

Log Revenue per Line

Log of telecommunications services revenues divided by number of main lines (i.e., revenues per line). Source: ITU.

Log GDP

Log of real GDP (billions of 1996 dollars), by country. Source: Summers & Heston.

Log Labor Force

Log of the total labor force of the country (millions). Source: Summers & Heston.

Log Geog. Area

Log of the geographic area of the country (thousands of square kilometers). Source: ITU.

Lag Penetration

Number of main lines per capita in the previous year, by country. Source: ITU.

Waiting List

Per capita waiting list for lines, i.e., number of people on the waiting list as a fraction of the total population, by country. Source: ITU.

Lag Log Telecom Stock

Log of the telecommunications capital stock in the previous year (log of $TECH_{t-1}$). Source: Summers & Heston.

Change in Digital Lines

Change in the percentage of digital lines, by country. Source: ITU.

Digital Dummy

Dummy variable for the presence of digital lines, by country. Source: ITU.

United States Data Set Variables

Change in Digital Lines

Change in percentage of total digital lines, by company. Source: DATAQUEST.

Total Lines

Number of total lines in the previous year, by company. Source: DATAQUEST.

Lag E/M Lines

The percentage of lines that were electromechanical in the previous year, calculated as lagged number of electromechanical lines divided by total lines, by company. Source: DATAQUEST.

Metro. Pop. (%)

Percent of population in a firm's service area living in metropolitan areas. Source: Statistical Abstracts of the United States. Service areas are from CCMI data and company web sites. The national populations were used for Other Independents.

Change in Metro. Pop.

The change in percentage of metropolitan population (i.e., the change in METPER). Source: calculated from METPER.

V. EMPIRICAL RESULTS

This section first highlights several stylized facts that shed light on the adoption process, then presents econometric analyses of the two data sets. The results for the U.S. data were developed in Shampine (2001) and are presented here for comparison with the international results.

A. International Results

Figure 1 tracks the percentage of digital lines for each country across time. Note that there is remarkable diversity in the initial deployment of digital technology in each country. Finland and France in 1980 were the first to do substantial upgrades with the U.S. a year behind.[8] Greece got the latest start in 1990. However, this graph hides several interesting dynamics since it does not capture the relative changes in analog and digital technology. Figure 1 by itself might suggest that once countries began to deploy digital switches that they did so exclusively and never looked back since it shows a uniformly increasing proportion of digital lines. That is not the case.

Figure 1.

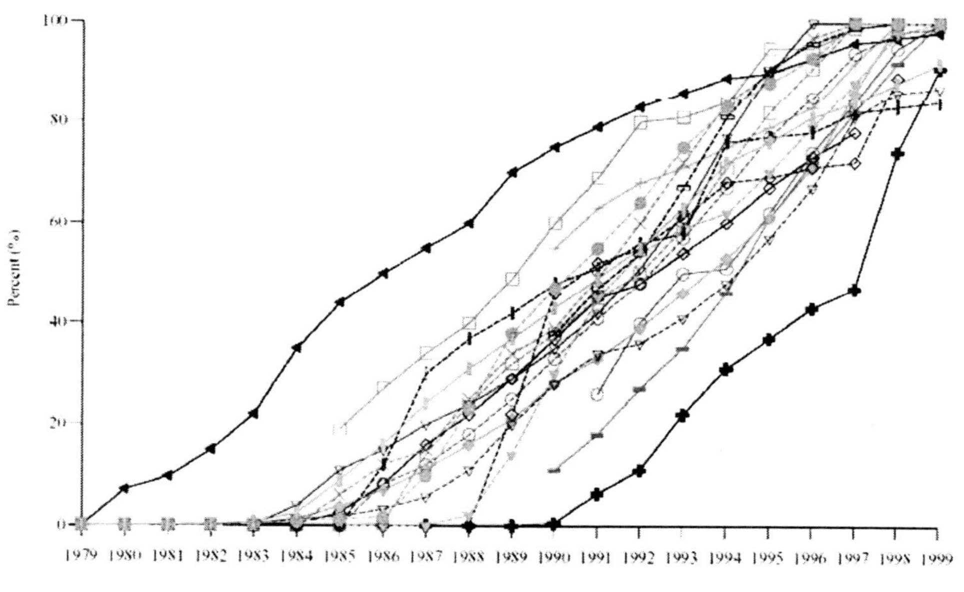

[8] France installed the first commercial digital switch in 1970, but digital lines did not comprise a meaningful portion of its infrastructure until 1980.

Table 2 presents a snapshot of the early stages of adoption for various countries. The columns provide the percentage of new lines that are analog or digital, except where a negative figure is presented. A negative number is the percentage of the installed base being upgraded to digital – that is, if the number of new digital lines exceeds the total number of new lines in the country, the table presents the portion of existing lines that is being upgraded. These figures are approximate as the data include only the total number of lines and the number of digital lines. To put it another way, the numbers in Table 2 assume that all digital lines added in a year are de novo installations unless the total number of additional digital lines exceeds the total number of new lines.[9] Table 2 presents the period from the first adoption of digital technology until a country discontinues deploying analog technology.

Digital technology does not immediately displace analog technology for most countries. Countries continued to deploy analog lines even after digital switches were available, and, in fact, often deployed both digital and analog lines at the same time. Some countries (Japan and the Netherlands) made a clean break with analog, completely discontinuing any new deployment of analog technology and replacing existing analog infrastructure with digital. Most countries, however, continued to deploy both technologies for years. In the case of Turkey, the technologies have continued to be deployed simultaneously for over a decade and the country has done little to replace existing analog infrastructure.

These data raise several questions. In particular, why do some countries begin adopting digital technology before others, and what determines the speed of the transition to digital? The first question is addressed by using a dummy for digital adoption as a dependent variable (0 if no digital adoption has occurred, 1 if some digital lines are present) and regressing various explanatory variables on it. The coefficients on the independent variables reflect their impact on the linear probability of a country adopting digital technology in a given year. Countries with a higher probability of adopting are thus likely to adopt earlier than other countries. Table 3 presents the results (excluding the coefficients on the country fixed effects).

The positive and significant coefficient on GDP indicates that wealthier countries are more likely to adopt digital technology than poorer countries. The coefficient on geographic area is negative and significant. The coefficient on government expenditures is not statistically significant.

[9] I also assume that all new lines are either analog or digital, not electromechanical. I am unaware of any new electromechanical switches being installed in these countries during this period.

Table 2. Analog vs. Digital Adoption

Country	Year	Analog Lines	Digital Lines
Belgium	1982	100%	0%
	1983	97%	3%
	1984	94%	6%
	1985	35%	65%
	1986	5%	95%
	1987	-4%	100%
Denmark	1982	100%	0%
	1983	86%	14%
	1984	56%	44%
	1985	23%	77%
	1986	0%	100%
	1987	-1%	100%
Finland	1979	100%	0%
	1980	98%	2%
	1981	100%	0%
	1982	98%	2%
	1983	100%	0%
	1984	4%	96%
	1985	-3%	100%
France	1979	100%	0%
	1980	42%	58%
	1981	68%	32%
	1982	29%	71%
	1983	-1%	100%
	1984	-9%	100%
Greece	1989	100%	0%
	1990	88%	12%
	1991	0%	100%
	1992	19%	81%
	1993	-6%	100%
Italy	1985	100%	0%
	1986	-4%	100%
	1987	8%	92%
	1988	-1%	100%
Japan	1984	100%	0%
	1985	-3%	100%
Netherlands	1988	100%	0%
	1989	-19%	100%
Norway	1985	100%	0%
	1986	95%	5%
	1987	-9%	100%
Portugal	1987	100%	0%
	1988	81%	19%
	1989	21%	79%
	1990	-9%	100%

Country	Year	Analog Lines	Digital Lines
Spain	1983	100%	0%
	1984	73%	27%
	1985	88%	12%
	1986	65%	35%
	1987	42%	58%
	1988	18%	82%
	1989	-3%	100%
Turkey	1985	100%	0%
	1986	35%	65%
	1987	17%	83%
	1988	42%	58%
	1989	33%	67%
	1990	16%	84%
	1991	33%	67%
	1992	10%	90%
	1993	35%	65%
	1994	-14%	100%
	1995	11%	89%
	1996	4%	96%
	1997	-1%	100%
	1998	1%	99%
	1999	0%	100%
United Kingdom	1982	100%	0%
	1983	94%	6%
	1984	82%	18%
	1985	86%	14%
	1986	83%	17%
	1987	-4%	100%
United States	1980	100%	0%
	1981	99%	1%
	1982	94%	6%
	1983	82%	18%
	1984	2%	99%
	1985	-4%	100%

Table 3. Regression Results: Probability of Adopting Digital Technology

Adj. R^2 = .6925
Observations = 363

Dependent Variable: Digital Dummy

Variable	*Parameter Estimate*	*T-Value*
Intercept	7.86	1.60
Log GDP	1.28	3.97
Log Geog. Area	-1.55	-5.45
Log Govt. Expenditures	-0.15	-0.88
Lag Penetration	0.04	0.49
Waiting List	-5.34	-3.33
Log Revenue per Line	-0.00005	-0.49
Lag Log Telecom Stock	0.44	5.17
Log Labor Force	-0.62	-1.30

Penetration and telecommunications capital are both measures of the installed base. Penetration reflects the ubiquity of service (lines per capita) while telecommunications capital measures the intensity of previous investment. The coefficient on lagged penetration is not statistically significant, while the coefficient on lagged telecommunications capital is positive and significant. That is, countries with larger investments in extensive telecommunications systems are more likely to adopt digital technology than countries with less extensive networks.

The demand variables offer mixed results. It is not clear what sign should be expected for revenue per line. Higher revenues may be indicative of higher demand, but they also may reflect constraints on supply. The coefficient here is not significant, however. The coefficient on the waiting list for lines is negative and significant. This variable is likely capturing the effects of previous deployment choices. Countries with large waiting lists have been slower than their counterparts in the past in deploying infrastructure. As discussed earlier, these choices are endogenous. These results suggest that countries which have been slow in the past choose to be slow with respect to digital as well. This is the opposite of the "leapfrog" effect found in the U.S. in Shampine (2001) where companies which had not invested in analog technology adopted digital systems more rapidly than those companies with large installed bases of analog systems. Finally, the coefficient on the size of the labor force variable is not statistically significant.

All of these results are for the probability of adopting in a given year. Table 4 presents regressions results concerning the rate of adoption. Here the dependent variable is the change in the percentage of digital lines. Again, country fixed effects are included in the specification but are not presented in the table.

Table 4. Regression Results: Change in Digital Lines

Adj. R^2 = .3559
Observations = 363

Dependent Variable:
Change in Digital Lines

Variable	*Parameter Estimate*	*T-Value*
Intercept	0.32	0.44
Log GDP	0.13	2.69
Log Geog. Area	-0.07	-1.67
Log Govt. Expenditures	-0.02	-0.91
Lag Penetration	-0.00	-0.26
Waiting List	-0.76	-3.19
Log Revenue per Line	.00002	1.07
Lag Log Telecom Stock	0.05	3.63
Log Labor Force	-0.19	-2.69

GDP is positively associated with the rate of diffusion, just as with the probability of adoption, and is statistically significant. Some authors have suggested that higher levels of adoption are causing higher levels of GDP, but even if true this is unlikely to be a contemporaneous effect. It is more likely within the same year that high levels of GDP are associated with high demand for telecommunications services.

The country size and the level of government expenditures are negatively associated with the rate of adoption. The coefficient on area is statistically significant at the 10 percent level, suggesting that countries with a larger geographic area are slower adopters of digital technology. This may be due to a negative correlation between country size and urbanization.[10] Alternatively, size may be correlated with other factors that are not accounted for here.

The two measures of installed base – penetration and telecommunications capital stock – have different signs. Lagged penetration is not significant, but the lagged telecommunications capital stock has a significant coefficient. As discussed below, this result is an interesting contrast to the U.S. results.

The coefficients on the demand measures are different than in the adoption regression in Table 3. The results for the waiting list for lines variable are the same. Countries which have larger waiting lists for lines, indicating that they have been relatively slow in deploying infrastructure in the past, are also slow in deploying digital infrastructure. However, the sign on the revenue per line variable is not statistically significant. The theoretical ambiguity of the sign may mean that different countries are experiencing different effects, resulting in an overall lack of significance. The sign on total labor force is negative and significant. This

[10] Population density is not significant when included as an explanatory variable, but, as previously discussed, overall population density is much less important than the distribution of density. Unfortunately, population density distribution data are not readily available for most countries.

result is puzzling. Most likely the size of the labor force is negatively correlated with per capita income (and thus demand), or is correlated with other excluded explanatory variables.

B. Comparison to U.S. Results

Table 5 presents the U.S. only results from Shampine (2001). These consist of two periods because a Chow test for structural change indicated that the two periods were significantly different. Company specific fixed effects were included in the specifications but are not reported here. The dependent variable is the change in the percentage of total digital lines, by company.

Table 5. Regression Results: Change in U.S. Company Digital Lines

Variable	1984 – 1989	1990 – 1996
Adj. R^2	.1725	.3066
Intercept	-0.417 (-0.47)	-2.303 (-1.95)
Lag Total Lines	0.00004 (2.84)	0.000007 (1.35)
Lag E/M Lines (%)	0.137 (2.56)	0.402 (5.48)
Metro. Pop. (%)	-0.272 (-0.20)	3.018 (1.83)
Change in Metro. Pop.	-1.353 (-0.90)	-2.298 (-2.00)

In the United States firms with larger networks (as measured by lagged number of lines) adopted more quickly, while firms with a larger proportion of analog switches (as measured by the lagged percentage of electromechanical lines) adopted more slowly. These measures are not directly comparable to the international results, but do present interesting contrasts. In the international results penetration which is the variable most similar to lagged total lines, has no significant effect, while total telecommunications capital stock is positive and significantly associated with faster and earlier adoption. The difficulty in comparing the results is that, in the U.S., both larger networks and more analog switches contribute to a larger capital stock, while the two have opposite effects on adoption. As a matter of theory, however, the effect of the analog/electromechanical choice on future adoption is ambiguous. The international results suggest that, regardless of specific choices, countries which have sunk a lot of money into telecommunications infrastructure in the past are likely to upgrade their systems sooner and faster than countries which have spent less. Either the analog/electromechanical choice effect varies between countries (always possible), is washed out by aggregation within the country (quite likely), or is overwhelmed by the overall significance of previous capital spending (possible but less likely).

In the U.S. it was possible to obtain line costs by technology and company type. However, they were not significant in any specification and so were excluded. The revenue per line measure for the international data also proved to be insignificant. These results are a bit surprising. They may be due to poor measurement (for example, the revenue variable may be missing some relevant services or failing to capture expected future revenues), or it may simply be that the difference between the benefits and costs is sufficiently large that relatively small movements in costs do not greatly affect adoption.

In the United States firms with larger proportions of urban residents (i.e., a more concentrated population) adopted more quickly than firms with more diffuse populations.[11] Oddly, firms with more rapidly growing urban populations actually adopted more slowly than their counterparts. In the international data, countries with a larger area adopted more slowly, although population density was not significant. Also, countries with larger labor forces adopted more slowly. Again, these measures are not directly comparable. However, all else equal, larger countries will tend to have the population split into more centers and those centers will tend to be more geographically diverse. This would make it relatively more expensive to obtain network externalities due to the necessity of upgrading the systems interconnecting each population center (the "long-haul" infrastructure). In the U.S. this long-haul infrastructure was largely provided by the interexchange carriers such as AT&T, who adopted digital technology rather rapidly. Accordingly, it was probably not a bottleneck in the U.S. for the local operating companies, but in other countries with a single company controlling both the long distance and local portions of the network, the issue is apparently more significant.

V. Conclusions

These results raise a number of issues. First, they provide some empirical evidence on the direction of effect for some theoretically ambiguous determinants of adoption. Internationally, GDP and telecommunications infrastructure are both associated with earlier and faster adoption. Geographic area, larger numbers of people waiting for lines to be installed and larger labor forces are associated with later and slower adoption. The U.S. results suggest that greater precision can be obtained by examining both the level and type of infrastructure. Otherwise, the results are similar except for one key difference.

These results suggest that the finding of leapfrogging within the U.S. does not apply to international comparisons. This is unfortunate since it supports concerns about the persistence of the "digital divide," where technological laggards remain permanently behind their neighbors (Campbell 2001). The results here suggest that countries which have been slow in the past will continue to be slow in the future, but offer little guidance as to how this might be remedied. The lack of leapfrogging implies, however, that a remedy is necessary. Countries will not spontaneously become rapid adopters.

The results also raise some concerns about previous studies on the relationship of telecommunications infrastructure and growth. In particular, there is significant variation in the types of infrastructure deployed. Unless the type of infrastructure is controlled for, analyses of the levels of infrastructure may provide misleading results. For example, Roller and Waverman (2001), have examined the 1975-1990 period, which was an early period in the history of digital adoption. Only one third of their observations have any digital deployment whatsoever, and the average level of digital lines is only 7 percent.[12] However, there is a significant increase in digital deployment among many countries at the end of their

[11] This is not the same as average population density. This is closer to a measure of urbanization. A country can have relatively high population density but still have a predominantly rural economy. It can also have a relatively low population density but have virtually all of the population concentrated in a few cities (such as Australia).

[12] These figures are obtained through independent analysis of Roller & Waverman's data set.

time period. Roller and Waverman use both penetration and capital measures but do not control for the type of technology deployed. Two countries, one of which purchases analog lines, another of which upgrades analog lines to digital lines, may spend the same amount of money, and the upgrading country will experience a smaller increase in penetration, but the digital upgrades may well have a greater effect on productivity than the analog lines. The results here indicate that contemporaneous GDP is strongly correlated with the timing and rate of digital adoption. This suggests that previous studies examining levels of infrastructure may be able to obtain greater statistical power by accounting for the type of infrastructure as well.

References

AT&T Broadband, "Investor Presentation," July 2001.

Berg, Sanford and Hamilton, Jacqueline. "Institutions and Telecommunications Performance in Africa: Stability, Governance, and Incentives," *Public Utility Resource Center working paper*, 2001.

Campbell, Duncan. "Can the digital divide be contained?" *International Labour Review*, January 1, 2001, v.140, iss.2, pp.119-141.

Doughert, P.; Germain, R.; and Droge, C. "Predicting EDI technology adoption in logistics management: the influence of context and structure; electronic data interchange," *The Logistics and Transportation Review*, 1995, iss.31, pp.309-373.

Farrell, J. and Saloner, G. "Installed base and compatibility: innovation, product preannouncements, and predation," *American Economic Review*, 1986, iss.76, pp.940-955.

Feder, G.; Just, R.; and Zilberman, D. "Adoption of agricultural innovations in developing countries: a survey," *Economic Development and Cultural Change*, 1985, iss.33, pp.255-298.

Greenstein, Shane; McMaster, S.; and Spiller, P. "The effect of incentive regulation on infrastructure modernization: local exchange companies' deployment of digital technology," *Journal of Economics and Management Strategy*, 1995, iss.4, pp.187-236.

Greenstein, Shane and Spiller, Pablo. "Modern telecommunications infrastructure and economic activity: an empirical investigation," *Industrial and Corporate Change*, 1995, iss.5, pp.647-665.

Gutierrez, Luis and Berg, Sanford. "Telecommunications Liberalization and Regulatory Governance: Lessons from Latin America," *Public Utility Resource Center working paper*, 2001.

Hardy, Andrew. "The Role of the Telephone in Economic Development," *Telecommunications Policy*, 1980, v.4, iss.4, pp.278-286.

Hausman, Jerry and Sidak, Gregory. "A Consumer-Welfare Approach to the Mandatory Unbundling of Telecommunications Networks," *Yale Law Journal* 1999: 417-505.

Joel Jr., A. "Digital switching – how it has developed," *IEEE Transactions on Communications*, 1979, iss. 27, pp.948-959.

Lichtenberg, Frank. "The Output Contributions of Computer Equipment and Personnel: A Firm-Level Analysis," *Economics of Innovation and New Technology*, 1995, vol. 3, pp.201-217.

Majundar, S. and Venkatarman, S. "Network effects and the adoption of new technology: evidence from the U.S. telecommunications industry," *Strategic Management Journal*, 1998, iss.19, pp.1045-1062.

Norton, Seth. "Transaction Costs, Telecommunications, and the Microeconomics of Macroeconomic Growth," *Economic Development and Cultural Change*, October 1992, v.41, iss.1, pp.175-196.

Rogers, E. *Diffusion of innovations*. The Free Press, New York, New York, 1995.

Roller, Lars-Hendrik and Waverman, Leonard. "The Impact of Telecommunications Infrastructure on Economic Development," in Peter Howitt, ed., *The implications of knowledge-based growth for micro-economic policies*. Calgary: University of Calgary Press, 1996, pp.363-387.

Roller, Lars-Hendrik and Waverman, Leonard. "Telecommunications Infrastructure and Economic Development: A Simultaneous Approach," *The American Economic Review*, September 2001, vol. 91, no. 4, pp.909-923.

Rose, N. and Joskow, P. "The diffusion of new technologies: evidence from the electric utility industry," The RAND Journal of Economics, 1990, iss.21, pp.354-373.

Rosenkranz, S. "Quality improvements and the incentive to leapfrog," *International Journal of Industrial Organization*, 1997, iss.15, pp.243-261.

Saloner, G. and Shepard, A. "Adoption of technologies with network effects: an empirical examination of the adoption of automated teller machines," *The RAND Journal of Economics*, 1995, iss.26, pp.479-501.

Shampine, Allan. "Determinants of the diffusion of U.S. digital telecommunications," *Journal of Evolutionary Economics*, 2001, iss.11, pp.249-261.

Theodosopoulos, N. and Chin, S. "Switching the Switches: Hiding in the Trojan Horse," *UBS Warburg, LLC*, September 21, 2001.

Travis, Paul. "Which way do we go? Options for upgrading analog switches." *Telephony*, 1989, iss.217, pp.32-36.

Witt, U. "'Lock-in' vs. 'critical masses' – industrial change under network externalities," *International Journal of Industrial Organization*, 1997, iss.15, pp.75

Chapter 3

NETWORK COMPETITION, TARIFF STRUCTURE, AND SUBSCRIBER BANDWAGONS: EVIDENCE FROM THE TAIWAN MOBILE TELEPHONE MARKET

W. Wayne Fu
School of Communication and Information
Nanyang Technological University

I. INTRODUCTION

The user of a communications product often emulates others' buying decisions (Rohlfs, 1974, 2001; Oren & Smith, 1981; Farrell & Saloner, 1985, 1986; Katz & Shapiro, 1985, 1986, 1994). To exploit benefits associated with network externalities (Economides, 1996; see Liebowitz & Margolis 1994 for a review), network shoppers optimize their choice-making contingent on which competing networks "early birds" had chosen. As such, they will be attracted to the network which has already installed a large base. The influence that early adopters have on later users' choices is referred to as "consumption bandwagon" (Rohlfs, 2001), which usually trademarks *durable* goods, such as telecommunications network services, whose value is determined by user size. Bandwagon effects frequently cause market domination by very few firms (Farrell & Saloner, 1986; Rohlfs, 2001).

In telecommunications, bandwagon behaviors can spring from various sources of externality, including the immediate financial value derived from network size. When choosing from interconnected network competitors, a subscription buyer is concerned with whether the selection minimizes the cost of network usage. That is, joining the most widely connected network reduces the possibility of costly interconnection. The process in which emerging network buyers over time mimic the pandemic choice sets off a *subscriber bandwagon*. This paper studies the subscriber bandwagon phenomenon in the Taiwan mobile phone market.

Over the last decade, cellular services have flourished worldwide. On this "mobile" wave, Taiwan has attained unprecedented adoption growth. In 1998 the government reinvented this mobile phone industry by opening up this previously state-monopolized sector to competition. Eight new cellular network licenses were issued to six start-ups. A fierce subscriber war ensued dealing the long stagnant market an explosive blow. In early 1998, only 6.86% of the Taiwanese had mobile connections. In barely three years, the penetration rate skyrocketed to 72.3% (Taiwan Directorate General of Telecommunications Statistics Report, 2001). This diffusion, illustrated in Figure 1, is unparalleled. However, behind the captivating mobile phone proliferation lies the fact that the subscriber market is very unequally divided among the competing network oligopolies. Networks with a larger subscriber base expanded markedly faster than others. Even casual observation can tell that a subscriber bandwagon played a role in the evolution of this mobile phone market.

Figure 1: The Growth of Mobile Phone Subscriptions in Taiwan

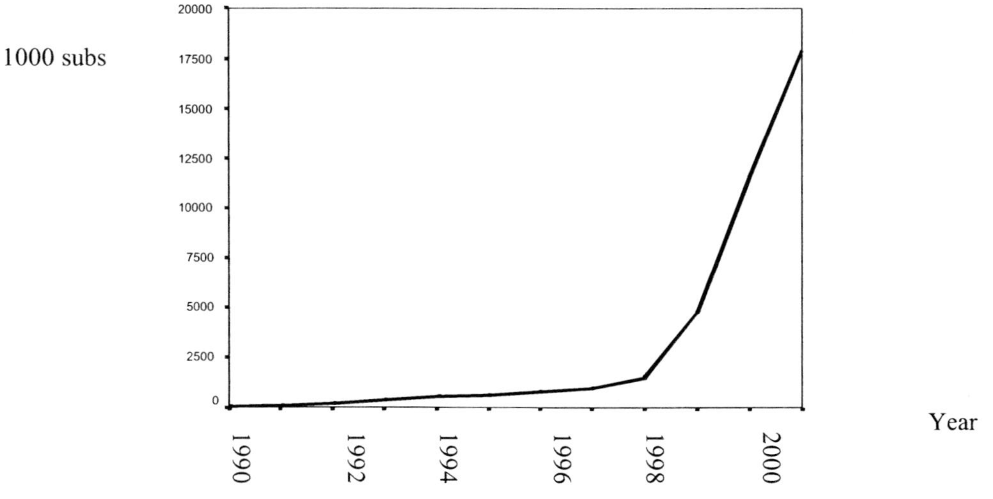

In this paper I construct an econometric model to measure subscriber bandwagons, and the factors which caused them, among the Taiwan mobile networks. The model reveals a tendency for new subscribers to flock to the most chosen networks—specifically, the number of subscriptions an operator obtains depends on its relative installed user base. Results suggest that the intensity of the subscriber bandwagon is related to subscribers' reaction to the destination-discriminatory pricing: the higher cost for calling outside the network impels users voluntarily to cluster their network affiliations. This investigation sheds light on the subscriber competition between interconnected networks, which is still lacking empirically. The main contribution resides on the empirical demonstration of interconnection fees as the raison d'être for the deepening network concentration or, even, domination seen in many telecommunications markets, a vexing issue to policymakers who desired competition.

The rest of this paper is organized as follows. Section 2 discusses network competition and reviews pertinent literature. Section 3 delineates Taiwan's mobile phone industry and its history. Section 4 provides the underpinning logic for the empirical approach used in the gauging of the bandwagon effect on network competition. Section 5 lays out the econometric model. Section 6 explains the data collected to estimate the model. Section 7 reports and interprets estimation results. Section 8 summarizes the key findings and considers policy implications.

II. INTERCONNECTION, CALL PRICING STRUCTURE, AND NETWORK COMPETITION

Research on network competition and interconnection has to date been preponderantly theoretical. Works on this area mostly address the access by competitive telecom retailers to a vertically-integrated incumbent's upstream network inputs (See Armstrong and Vickers 1996, 1998 for reviews on theories). However, the competition between mobile carriers is a different situation (Armstrong, 1997): each carrier vies for subscribers to run its own network but, still, is interconnected with all other networks in order to provide inter-network end-to-end connectivity.

Networks levy a fee for terminating calls originating from different networks. This interconnection fee raises operators' cost for servicing outbound calls (those completed *outside* the originating network) relative to internal calls (those initiated and terminated in the same network). Serious implications arise when this charge exceeds the actual cost of termination. A caller-pays pricing regime based on this cost structure will have to tariff outbound traffic considerably higher than internal ones. Consequently, subscribers to different networks incur dissimilar costs for reaching the very same person or a given group of persons.

Economides, Lopomo, and Woroch (1996) warned against the potential of such tariff schemes to tilt network competition toward market domination. They recognized a *pecuniary* externality that is effectuated by the destination-discriminatory pricing scheme: a network with more users is of higher subscription value because it allows one to call more people internally and *cheaply*. The pecuniary externality essentially impels prospective buyers toward a network with dominant coverage *ceteris paribus*, and accordingly, instigates a subscriber bandwagon. Pricing strategies that implement this destination discrimination are widespread. Examples are MCI's *Friends-and-Family* and AT&T's $19.95 unlimited internal long distance calls.

The Hotelling-differentiation model of Laffont, Rey, and Tirole (1998) demonstrated the incentive of competition network duopolists to collude to maintain high interconnection charges in order to soften the rivalry in the retail user market. High interconnection prices diminish the substitutability of subscriptions to them. This result was also found by Armstrong (1998). In fact, it is shown that the removal of interconnection fees reduces consumption inefficiency on account of price markups, assuming that a call benefits equally the sender and recipient (DeGraba, 2000). Economides and Himmelberg (1995) made clear that even competition cannot ensure efficiency optimality since the benefit of network externalities is not internalized by individual sellers.

Economides (1996) surveyed competition of network products and elucidated the path-dependent evolution of network markets: historic events amount to the self-fulfilling expectation of network consumers about the future of the market. The path-dependence was illustrated by Katz and Shapiro's (1986) brisk remark—"the relative attractiveness today of rival technologies is influenced by their sales histories."

Some studies have scrutinized mobile phone competition *per se*: Valletti (1999) regarded service area coverage as a differentiated vertical quality. He proved that, in equilibrium, mobile network adversaries supply different coverage to alleviate price wars. Doyle and Smith (1998) advocated the receiver-pays principle as one regulatory remedy for the limited mobile competition in Britain. They argued this pricing scheme could reduce relatively high calling rates and then boost market demand. However, these studies did not account for network effects.

III. THE TAIWAN MOBILE TELEPHONE MARKET

Cellular telephone service was introduced to Taiwan in 1990. Until early 1999, this service had been provided solely by the Directorate General of Telecommunications (DGT) using AMPS. The DGT, under the supervision of the Ministry of Transportation and Communication, was a government agency that operated telecommunications services and also oversaw these sectors. In 1996, as a milestone action by the Taiwan government to liberalize telecommunications and to set the stage for the coming competition, DGT was restructured to become an independent regulatory authority—all its service segments were privatized and incorporated into a publicly-traded company, Chunghua Telecom.

In 1998, the DGT held a competitive application for allocating eight new GSM licenses—two nationwide (1800 MHz) and six regional (900 and 1800 MHz each in the north, central, and south regions, respectively). Fifteen contestants participated in this beauty pageant and most of them were collaborations of conglomerate investors, domestic and international. Among them, Taiwan Cellular obtained a nationwide permit, Far Eastone a nationwide and a north, KG Telecom a north, Tuntex a central and south, TransAsia a south, and MobiTai a central. These newcomers all started their operation in early 1999 rivaling the incumbent Chunghua Telecom.

In the market share battle, these operators frequently mobilized marketing campaigns to lure subscribers. On-net call rate discounts, subscription referral rewards, and lowered installation fees for multi-account subscriptions were commonly seen. The ferocious competition kept the players watchful on one another's price offers. Initially, Chunghua Telecom was the price leader, with Taiwan Cellular later taking over the leadership, but any changes in their tariffs were swiftly matched by the others.

At the core of the fledgling operators' market survival was their access to Chunghua's networks, both wireless and wireline. Pursuant to the 1999 Telecommunications Act, the DGT promulgated rules to regulate all telecom carriers' interconnection– *the Administrative Regulation Governing Interconnection of Type 1 Telecommunications Enterprises*. This order forbade telecommunications carriers to refuse requests for interconnection and mandated the principles to guide the bilateral negotiations between carriers to set interconnection rates and terms. In the ruling the interconnector is required to unbundle its network in order to allow access to elements required for interconnection. The interconnection charge was to be based on the actual cost of the attributing elements as calculated by the TELRIC (total element long-run incremental cost) method (the method developed by the US Federal Communication Commission). TELRIC was viewed as an entrant-favoring pricing method and its use was intended to foster competition.

Asymmetric burdens were imposed on Chunghua Telecom to prevent it from exploiting its incumbent position. The DGT requires the dominant carrier to file and announce any new tariff plan in 14 days' advance of effect. This afforded the entrants more preparation and nimbleness in reacting to the market situation and the price war waged by the incumbent.

Despite the use of TELRIC to calculate interconnection cost, actual interconnection fees have been remarkably high. The average floor in Taiwan was NT$1.6 (US$0.053)/min while the rates in Hong Kong and Singapore were merely NT$0.27 (US$0.009)/min (Chou, 2000). Possible reasons are manifold. Competition and interconnection were epochal telecommunications events to Taiwan. Neither regulators nor firms had the experience and knowledge to execute precisely the relatively complicated TELRIC procedure. Thus, interconnection negotiations were frequently made using rules of thumb.

Companies have often complained about Chunghua Telecom's high interconnection charge and its reluctance to decrease it. Although the marketwide fee level has dropped over these past years, the decrease has not been sufficient to approach the competitive ideal. Moreover, even some other operators who once opposed the high interconnection cost seem to have become comfortable keeping this charge high after they secured some critical size of users. These companies have the same incentive as Chunghua Telecom: to raise other networks' (in particular, small ones) cost for competing in the retail subscription market.[1]

The Interconnection Regulation mandates a caller-pays regime for all telecommunications pricing—the caller bears all costs incurred (inclusive of the interconnection cost if applied) and the call recipient pays nothing irrespective of the call origination and termination. Tariffs of the operators typically differentiate between within-network and cross-network calls and designate disparate rates for them. Outbound rates are considerably marked up on top of the interconnection cost. Normally, cross calls (eg. NT$6–7.2/min or US$0.2–0.24/min) were priced at twice the level of within calls (eg. NT$3/min or US$0.1/min). This rate discrimination as to destination makes one's calling behavior more sensitive to the operator's costs and so makes callers more conscious about choosing the network to subscribe.

Another factor affecting the Taiwan mobile industry is service coverage. Initially, two companies (Taiwan Cellular and Far Eastone) in addition to the incumbent were licensed to serve the entire country market under one permit. KG Telecom joined the all-country club in January 1999 through acquiring Tuntex. Contrasted with the nationwide players whose subscriberships soared quickly, the single-region operators– TransAsia and MobiTai– encountered difficulty in attracting sign-ups.

First, scale economies in marketing, billing, distribution, and the like are hard for single-region operators to materialize provided the limited market base. Another problem for regional operators is that because of their small subscribership, users, unavoidably, are forced to call outward to reach anyone outside their region or subscribers of other networks within their region. Of course, small area coverage also circumscribes users' mobility. To overcome the latter problem, operators from different regions formed alliances to enable their subscribers to roam across areas. But, calls made during roaming still incur a full interconnection charge and are priced by the cross rates.

Notwithstanding their concurrent births, the new operators took different paths. In May 1998, the monthly signed subscriptions were split as follows: Taiwan Cellular 32.1%, Far Eastone 14.7%, KG Telecom 12.8%, MobiTai 4.9%, TransAsia 3.3%, and Tuntex 1.8%, with the incumbent– Chunghua Telecom 30.4%. The market share imbalance had increased by September 2000: Taiwan Cellular 43.1%, Far Eastone 16.7%, KG Telecom 10.4%, MobiTai 3.8%, and TransAsia 2.1%, plus Chunghua Telecom 23.9%.[2] The Herfindahl-Hirschman

[1] Katz and Shapiro (1994) argued that a dominant network benefits from reducing cross-system compatibility to undermine its competitors' appeal. Besen and Farrell (1994) made similar points. Bental and Spiegel's (1995) model of network competition also attested that.

[2] In the interim Tuntex was merged into KG Telecom.

Index (HHI) of the share distribution rose from 2373 to 2835, correspondingly. If not considering Chunghua, we see an even stronger increase in share concentration: the HHI goes from 2992 to 3909. For an adequate account of the concentration transition and the subscription bandwagon, I focus on the newcomers, whose situations are similar but different from that of Chunghua. Asymmetric regulatory treatments and inherited bureaucratic inertia shackled Chunghua and cost it its leadership in the mobile subscription race (Chou & Liu, 2002).

The next section outlines the approach used to examine the subscriber bandwagon.

IV. Assessing the Subscriber Bandwagon

Previous bandwagon studies relied on "macro" empiricism—inferences on individuals' consumption were drawn from observation of the aggregate (market) behavior (Greenstein, 1993). Within this approach, the bandwagon effect is validated if the number of agents using a product is empirically related to total past adoptions (Levin, Levin, & Meisel, 1987). Here I also conduct a macro-level analysis. The micro alternative – to discern individual subscribers' varying preference for unequally-sized networks – was infeasible in that it requires a formidable amount of individual-level information, mostly inaccessible (and unobservable). For this mobile network context, we examine the proposition: the division of new subscriptions among different networks is related to each relative installed base.

To test this proposition, we need to characterize the competition between the Taiwan mobile networks and network adoption by individual nonsubscribers. Without doubt, the network subscription is a durable good. There is little dispersion in calling rates across the sellers, and the market prices decline over time because of technological innovation and increasing cost savings from scale economies. Section 5 compares monthly minute rates charged by the different operators.

The following assumptions are made to reduce the complexity of the examination of subscriber bandwagons. Facing the competitive network supply, the subscription decision of an individual person is equivalently one that involves the choice of adoption time and the choice of which network to join. It is assumed that the person will begin the mobile connection at a time when mobile subscription yields a net positive benefit, that is a function of the current calling rate, the market price of hand sets, the number of countrywide subscribers, and his or her private valuation characteristics. Once determined to adopt, the person then will choose from the competing networks. The bandwagon conjecture states that it is more probable for the person to enter a network that holds a bigger subscriber inventory. A bigger network is anticipated to be the one that has already connected to more people whom the subscriber will possibly call. By using the same network the expensive off-net connection can be eschewed.

These characterizations of the market, generally aligning with industry wisdoms, give pragmatic advantages in assessing subscriber bandwagon. With them, we can simply use shares of current sold subscriptions as an indicator of how competing networks are valued by market consumers without specifying personal attributes.

V. ECONOMETRIC SPECIFICATIONS AND VARIABLES

This section sets forth the econometric model for assessing the network effect on Taiwan mobile subscription sales. These mobile operators are interconnected and price-discriminate between call destinations. The goal is to test the hypothesis that in this environment a larger network base attracts a larger share of new accounts.

We track each firm's current market subscriptions share. $MS_{i,t}$ represents the fraction of all accounts sold during month t which are from network i. This sales share is a function of i's relative marketwide installed base by time t–1 ($TS_{i,t\text{-}1}$), the termination-based pricing ($Rate\Delta_t$), and firm-specific features, $\mathbf{Z}_{i,t}$, as shown by equation (1). Variable $TS_{i,t\text{-}1}$ is i's cumulative penetration among existing users, and is postulated to exert a bandwagon influence on $MS_{i,t}$.[3] The underlying relationship between $MS_{i,t}$ and $TS_{i,t\text{-}1}$ is specified by a logistic function $MS(TS) = \dfrac{1}{1+\left(e^{1-\delta\cdot TS}\right)^{\alpha}}$ for $0 < TS < 1$. This function has the tractability needed for estimation. $MS(\cdot)$ increases roughly from 0 monotonically, initially in convexity and then in concavity, in TS. Parameter δ determines the value of TS by which the switch in the curvature takes place. How this curvature changes depends on parameter α. When α increases from 1, the correspondence of TS to $MS(\cdot)$ becomes polarized: large (small) TS leads to a disproportionately larger (smaller) value of $MS(\cdot)$. Accordingly, the greater α is, the more skewed the $MS_{i,t}$ distribution is among i relative to the distribution *ceteris paribus*. For α less than 1, the $MS_{i,t}$-$TS_{i,t\text{-}1}$ relation is more linear. So, parameter α indicates the existence of a subscriber bandwagon and is called the *bandwagon* index for convenience.

$Rate\Delta_t$ is the average of the external vs. internal rate differential for all i at time t. This variable is factored into α to reveal any effect of destination-sensitive pricing: $\alpha = \alpha_0 + \alpha_R\, Rate\Delta_t$. If destination-sensitive tariffs dispose people towards the most connected network in a bandwagon, higher $Rate\Delta_t$ should raise α, i.e. a positive sign on α_R. In other words, the decreasing rate-discrimination between on- and off-net calls will neutralize the propensity of account purchasers to form into the bandwagon.

$Rate\Delta_t$ is endogenous to $MS_{i,t}$. The price markup of off-net over on-net traffic, inevitably and largely, is ascribed to the imposition of an interconnection fee by the inter-network traffic terminating firm over and above actual operation costs. The level of the interconnection fee is negotiated between any two interconnected networks. To gain an advantage against competitors in selling subscriptions, an operator has a strategic incentive to set the level of the interconnection fee to his favor. The optimal level is, at least partially, dependent on a firm's subscriber share (Laffont & Tirole, 2000, p. 9). For this reason, $Rate\Delta_t$ should be taken as simultaneously determined with the market share variables.

$$MS_{i,t} = \frac{1}{1+\left(e^{1-\delta\cdot TS_{i,t-1}}\right)^{\alpha}\cdot g\left(\mathbf{Z}_{i,t}\right)} \qquad (1)$$

[3] What influences the choice of subscribers starting at time t is market shares of time t – 1, rather than t. A user joining a network during period t will not be listed until the beginning of time t + 1. Also, it is assumed that individuals have the ability to make informed decisions about network bases.

where $\alpha = \alpha_0 + \alpha_R\, Rate\Delta_t$ and $\mathbf{Z}_{i,t} = (Promo_{i,t}, Nationwide_{i,t}, TwoRegion_{i,t})$
Equation (1) can be reconfigured as (2).

$$\frac{1}{MS_{i,t}} - 1 = \left(e^{1-\delta \cdot TS_{i,t-1}}\right)^{\alpha} \cdot g(\mathbf{Z}_{i,t}) \tag{2}$$

Taking logs and having $Y_{i,t}$ to be $\frac{1}{MS_{i,t}} - 1$, we have expression:

$$\log Y_{i,t} = \alpha(1 - \delta \cdot TS_{i,t-1}) + \log g(\mathbf{Z}_{i,t})$$

Then, we plug $\alpha_0 + \alpha_R\, Rate\Delta_t$ into α:

$$\log Y_{i,t} = \alpha_0 + \alpha_R Rate\Delta_t - \alpha_A \delta \cdot TS_{i,t-1} - \alpha_R \delta \cdot Rate\Delta_t \cdot TS_{i,t-1} + \log g(\mathbf{Z}_{i,t}) \tag{3}$$

Then we specify $\mathbf{Z}_{i,t}$. Variables of $\mathbf{Z}_{i,t}$ are included to control variations in $MS_{i,t}$. First, is a dummy variable that indicates promotional/marketing effort. $Promo_{i,t}$ is equal to 1 if operator i offered any promotional deals at time t, and 0 otherwise. In the competition for market share, operators heavily employed aggressive campaigns to solicit subscriptions. Rate discounts, account-handset bundles, and gifts were very common.

$Nationwide_{i,t}$ and $TwoRegion_{i,t}$, also dichotomous variables, name whether network i at time t operates a nationwide (3-region) or 2-region coverage, respectively, compared against single-region operators. Service coverage area is valued as a quality to mobile users (Valletti, 1999). As mentioned, though all operators allow subscribers to stay connected countrywide either under their own coverage or through a roaming alliance, a user of a network with a limited service area has to pay off-net rates for calls made during cross-region roaming. Off-net rates are charged since the network which handles the roamer' calls bills the home network for this interconnection.

Assuming a particular function for $g(\mathbf{Z}_{i,t})$ in (3), we have the regressible equation (4) with $\beta_1 = \alpha_R$, $\beta_2 = -\alpha_A\delta$, and $\beta_3 = -\alpha_R\delta$. Also, this estimation uses the least squares dummy variables (LSDV) model. Network dummies (d_i) are introduced as respective intercepts and will explain fixed effects caused by individual networks' idiosyncrasies. In estimating (4), a panel dataset on the six Taiwan mobile phone companies was compiled. The six are Taiwan Cellular ($i = 1$), Far Eastone ($i = 2$), KG Telecom ($i = 3$), TransAsia ($i = 4$), MobiTai ($i = 5$), and Tuntex ($i = 6$).

$$\log Y_{i,t} = \sum \gamma_i d_i + \beta_1 Rate\Delta_t + \beta_2 TS_{i,t-1} + \beta_3 Rate\Delta_t TS_{i,t-1} + \lambda_1 Promo_{i,t} + \\ + \lambda_2 Nationwide_{i,t} + \lambda_3 TwoRegion_{i,t} + \varepsilon_{i,t} \tag{4}$$

VI. Data

The data were taken primarily from the Telecom Statistics Reports (Taiwan Directorate General of Telecommunications, 2000 & 2001) and individual mobile service firms' trade publications. The first were comprehensive descriptions, both factual and statistical, of the Taiwan telecommunications status quo. These reports, updated quarterly, systematically profiled the mobile operators and their subscriberships. The phone companies, on their websites and in their corporate reports, publicized tariffs in detail and chronicled, for the most part, their subscribership growth. Information recorded from different sources was cross-checked to ascertain the reliability of the subscriber numbers being entered. Since only cumulative numbers regarding the subscribership were available, $MS_{i,t}$ was calculated simply by subtracting $TS_{i,t-1}$ from $TS_{i,t}$. The six-firm dataset covers a duration of 40 months (for $i = 1$ to 5, $t = 1$ to 40; for $i = 6$ Tuntex, $t = 1$ to 10) from March 1998 through June 2001.

All networks price-discriminated between on and off-net calls and offered multiple packages of two-part tariffs. One subscription package stipulates a particular combination of a fixed monthly fee and on and off-net calling rates. As a departure from the standard two-part tariff, these pricing schedules did not designate any call-minute allowance, but, instead, let a customer's fixed subscription fee be deductible from the call charge incurred in a month. (That is to say, the user pays nothing other than the subscription fee if his or her bill is smaller; otherwise, the actual bill.) It was found that one network's tariff menu was fairly akin to another's. Given more than one rate used by each operator, to have a consistent comparison benchmark, the measurement of the rate variable ($Rate\Delta_t$) chooses the package with the cheapest subscription fee within an operator's menu at a given time. Minute-rates (internal and external) specified in this tariff were used to calculate $Rate\Delta_t$. (To see if the estimation results were peculiar to this measurement choice, the most expensive, or say, "luxurious," package was also used. The econometric results did not show any significant qualitative deviation.) The claim that these operators' calling rates were in close parity finds support. The standard deviation of all operators' internal (external) minute-rates at a given time is, steadily, far smaller than the mean, with the ratio ranging from 1:13.96 to 19.54 (from 1:15.07 to 19.82) throughout the data duration. And, rates show no systematic association with a network's subscriber size. Unsurprisingly, the internal/external calling rate differentials of individual operators were very similar at all times. Moreover, $Rate\Delta_t$ declined over time, which may be explained by regulatory pressure for ever-lower interconnection charges.

National newspaper and television network broadcasts were periodically searched for mobile service subscription advertising/commercials. A subscription promotion ($Promo_{i,t}$) was identified when a network circulated through these mass media any messages of its subscription offers with lower rates than regular. Last, the data on $Nationwide_{i,t}$ and $TwoRegion_{i,t}$ were drawn from companies' publicized information. A few network licensees' coverage altered during the data period because of consolidation or acquisition.

VII. Results and Discussion

Significant autocorrelation is unveiled from (4) both by the Ordinary Least Squares (OLS) residuals and the Durbin-Watson statistic. This autocorrelation and the endogenity of (4) require a consistent and efficient estimator. The modified Feasible Generalized Least

Squares (FGLS) procedure specified by Greene (2000, p. 689) is used for (4), assumed with first-order autoregression AR(1).

The modified FGLS estimation begins with assessing the autocorrelation coefficient of equation (4). This is done by running a Two-Stage Least Squares (2SLS) regression for (4). In the 2SLS, the two-period lagged values are used as the instrument for the endogenous variable, *Rate*Δ_t, in the first-stage regression. The autocorrelation coefficient $\hat{\rho} = \dfrac{\sum \varepsilon_{i,t} \cdot \varepsilon_{i,t-1}}{\sum \varepsilon_{i,t}^{\ 2}}$ is computed from residuals, $\varepsilon_{i,t}$, of this 2SLS regression. Then, equation (4) is re-estimated, by Ordinary Least Squares (OLS) but with the transformed data: the 2SLS predicted value is entered for *Rate*Δ_t and, for each of all other terms (the dependent and regressors), say $k_{i,t}$, the new value $k_{i,t} - \hat{\rho}k_{i,t-1}$ is entered. The results obtained from this regression give the final estimates of (4), i.e. those in Table 1.

Table1: The Regression of logY_i, AR(1) FGLS

	Model 1	Model 2	Model 3
γ_1	– 0.081		
	(0.062)		
Γ_2	– 0.051 *		
	(0.021)		
γ_3	0.036		
	(0.077)		
γ_4	0.022 *		
	(0.010)		
γ_5	0.061 *		
	(0.019)		
γ_6	0.070 *		
	(0.032)		
β_1	1.167 *	1.547 *	1.692 *
	(0.066)	(0.120)	(0.274)
β_2	– 2.845 *	– 3.163 *	– 3.783 *
	(1.031)	(0.945)	(1.005)
β_3	– 3.799 *	– 3.692 *	– 3.604 *
	(0.984)	(0.660)	(0.413)
λ_1	--1.517*		– 2.524 *
	(0.395)		(0.446)
λ_2	– 0.635 *	– 0.951 *	
	(0.171)	(0.218)	
	– 0.324 *	– 0.508 *	
λ_3	(0.144)	(0.233)	

Note: (Standard errors in brackets)
* Significant at 95% confidence level

Table 1 presents the regression results. All regressors' coefficients retain statistical significance. Parameters α_0, α_R, and δ can be recovered from these coefficients. The rate-affected bandwagon index α_R has the value of 1.167 derived from $\alpha_R = \beta_1$ and the rate-unrelated index α_0 is 0.874 because $\alpha_0 = \frac{\beta_1 \cdot \beta_2}{\beta_3} = \frac{1.167(-2.845)}{-3.799}$. Parameter δ is calculated as it equals $-\frac{\beta_3}{\beta_1} = -\frac{-3.799}{1.167} = 3.255$. According to restricted tests the estimates of α_0, α_R, and δ all are significantly above zero.

The fairly large value of the index α_0 (tariff-invariant) substantiates the fact that large values of $TS_{i,t-1}$ cause even larger values to $MS_{i,t}$. That is, the subscription market had experienced persistent "tilting"—big networks tended to be sought after the most among new mobile users. Account buyers emerging at a given time mostly subscribed to the most popular networks. Actually, $MS_{i,t}$ perpetuates an accumulation of continual effects: any subscription intake from a bandwagon "snowballs," providing an already large base with an even bigger sale boost and thus another cycle of bandwagon reinforcement. That is, a subscriber bandwagon feeds itself through positive repetition: $TS_{i,t} = TS_{i,t-1} + MS_{i,t}$. The positive feedback reiterates between share gaining and the perceived network value, accruing to the winning network's growth. In contrast, small networks are haunted by a downward spiral—their market shares become smaller and smaller. Hence, any disparity in the share of subscriptions will worsen over time. As Armstrong (1997, p. 67) vividly recounts: "a firm supplying its phone service over a smaller network would be placed at a severe competitive disadvantage, in that all else being equal it would have to charge a lower price for the service to counteract the lower utility from being able to call a small number of other people." It is interesting that the smaller networks did not consistently charge significantly lower prices, since by charging rates similar to their larger competitors they contributed to their downward spiral (although see below for the effect of promotions).

The significant and positive α_R further adds to the value of the composite bandwagon index α. When $Rate\Delta_t$ is higher, the subscription bandwagon is intensified. That is to say, the destination-based price discrimination exacerbates the polarization of different-sized networks' growth path. This specific parametric result buttresses the view that different calling costs for reaching inwards and outwards "beautify" the network that has been established widely in the retail subscription market. The pecuniary externality of network users' adoption behavior is attested. Figure 2 plots $MS_{i,t}$ vs. $TS_{i,t-1}$ based on the estimated values of α_0, α_R , and δ, and the mean of $Rate\Delta_t$ (= 3.110 dollars/min.), when leaving alone $g(\mathbf{Z}_{i,t})$. It is clear that operators who cannot hold roughly 20% about of total marketwide subscribers are unable to keep the same share level in attracting new accounts. On the other hand, those who manage to reach this threshold will get even more.

Figure 2: The Relationship of $MS_{i,t}$ and $TS_{i,t-1}$ with estimated $\alpha = 4.503$ and $\delta = 3.255$

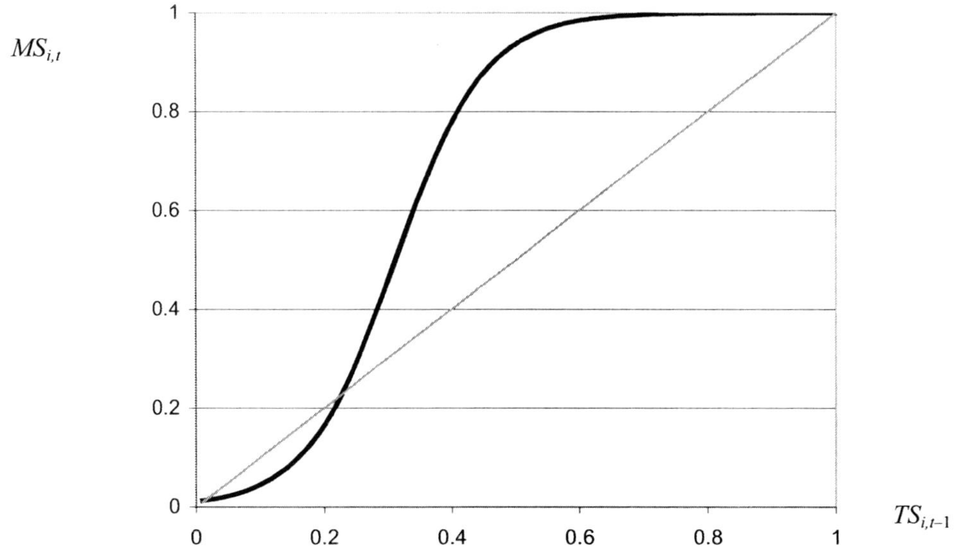

$Promo_{i,t}$ significantly decreases $\log Y_{i,t}$ (increases $MS_{i,t}$). It means that marketing efforts indeed pay off in obtaining subscriber shares. This result coincides with anecdotal industry perceptions. In particular, smaller operators may use promotions (price decreases) to attempt to offset their disadvantage with respect to network externalities.

$Nationwide_{i,t}$ and $TwoRegion_{i,t}$ both have a significant and negative association with $\log Y_{i,t}$. Multi-region licensees acquire considerably larger monthly shares relative to single-region operators.

In order to ensure the reliability of the model, several specification variants (using different sets of the regressors) are used. These results are shown in Table 1 as well and demonstrate that the coefficients are stable and consistent. This specification stability helps dispel the risk that the findings arise simply by chance or from any other unidentified dynamics.

VIII. Concluding Remarks

Taiwan had embraced the pro-market paradigm to spur telecommunications growth. Competition was ushered into not only the mobile service market (both 2G and 3G), but the fixed-line sector. Benefits of competition such as unleashed market demand, lower rates, and more service innovation have resulted. Despite these widely expected benefits of competition, empirical understanding of network subscriber competition remains fairly crude.

The existence of rivalry in the retail subscription market does not necessarily ensure the supply of interconnected telecommunications services will be, or be close to, ideally competitive. A carrier can compete with others for subscribers but still be a monopoly serving the users already on its own network, in the presence of sufficiently high switching costs.

This fact makes it difficult for interconnection between different networks to be supplied as competitively as efficiency requires. Highly-priced interconnection causes incompatibility between different networks and so encourages consumption bandwagons.

This study demonstrates the effects of network externalities on the Taiwan mobile market. Bandwagon behavior is found among the mobile phone subscribers. Results also show that asymmetric costs for connecting to varying-size networks significantly increase the subscriber bandwagon. The subscriber bandwagon takes place in the way that networks with different sizes at any given time divide all new subscribers disproportionately. The large networks devour the much larger shares of current sign-ups and increasingly marginalize the small ones, offset only by price reductions from the small firms. This self-reinforcing bandwagon process gives rise to domination by only a few networks over the subscriber market. Katz and Shapiro (1994) acknowledged: "Because of the strong positive feedback elements, systems markets are especially prone to 'tipping,' which is the tendency of one system to pull away from its rivals in popularity once it has gained an initial edge."

Given this, it is crucial to maintain a level playing field for network industries. The regulator should not downplay the influence of any "initial gain," even small, that results from an anticompetitive exercise. Circumspection should be used to prevent any unjust market advantage from being parlayed into a large, lasting one.

The findings also demonstrate that the pricing scheme which charges more for external calls significantly amplifies subscribers' preference for a larger network. For one, attaching to a large network market gives immediate financial savings. It is quite natural that outbound network transmission is more expensive simply because of the extra interconnection cost. However, dominant carriers may set the interconnection fee too high in order to cripple small competitors' ability to attract subscribers, in which case action by policymakers is warranted.

REFERENCES

Armstrong, M. (1997). Competition in telecommunications. *Oxford Review of Economic Policy*, 13(1), 64-82.

Armstrong, M. (1998). Network interconnection in telecommunications. *The Economic Journal*, 108(448), 545-564.

Armstrong, M. & Vickers, J. (1996). The access pricing problem: a synthesis. *Journal of Industrial Economics*, 44(2), 131-150.

Armstrong, M. & Vickers, J. (1998). The access pricing problem with deregulation: a note. *Journal of Industrial Economics*, 46(1), 115-121.

Bental, B. & Spiegel, M. (1995). Network competition, product quality and market coverage in the presence of network externalities. *Journal of Industrial Economics*, 43(2), 197-208.

Besen, S.M. & Farrell, J. (1994). Choosing how to compete: strategies and tactics in standardization. *Journal of Economic Perspectives*, 8(2), 117-131.

Chou, G.W. (2000). *The legend of the Taiwan Cellular* (Chinese). Taipei, Taiwan: Zhan Zhi Enterprise Co.

Chou, Y.T. & Liu, K.C. (2002). The Paradoxical Impact of Asymmetric Regulation in Taiwan's Telecommunications Industry: Restrictions and Rent-seeking. *Telecommunications Policy*, forthcoming.

DeGraba, P. (2000). Efficient interconnection regimes for competing networks. The 2000th Telecommunications Policy Research Conference.

Doyle, C. & Smith, J.C. (1998). Market structure in mobile telecoms: qualified indirect access and the receiver pays principle. *Information Economics and Policy*, 19(4), 471-488.

Economides, N., & Himmelberg, C. (1995). Critical Mass and Network Evolution in Telecommunications. *Toward a Competitive Telecommunication Industry, Selected Papers from the 1994 Telecommunications Policy Research Conference*. G. Brock, ed., Mahwah, NJ: Lawrence Erlbaum Associates.

Economides, N. (1996). The economics of networks. *International Journal of Industrial Organization*, 14(6), 673-699.

Economides, N., Lopomo, G., & Woroch, G. (1996). Regulatory pricing policies to neutralize network dominance. *Industrial and Corporate Change*, 5(4), 1013-1028.

Glynn, S. (1994). How many cellular licences should there be? The economic feasibility. *Telecommunications Policy*, 18(2), 91-96.

Greene, W.H. (2000). *Econometric Analysis*, 4th ed. Upper Saddle River, NJ: Prentice Hall.

Greenstein, S.M. (1993). Did installed base give an incumbent any (measurable) advantages in federal computer procurement? *RAND Journal of Economics*, 24(1), 19-39.

Katz, M.L. & Shapiro, C. (1985). Network externalities, competition, and compatibility. *American Economic Review*, 75(3), 424-440.

Katz, M.L. & Shapiro, C. (1986). Technology adoption in the presence of network externalities. *Journal of Political Economy*, 94(4), 822-841.

Katz, M.L. & Shapiro, C. (1994). Systems competition and network effects. *Journal of Economic Perspectives*, 8(2), 93-115.

Laffont, J.J., Rey, P., & Tirole, J. (1998). Network competition: I. Overview and nondiscriminatory pricing. *The RAND Journal of Economics*, 29(1), 1-38.

Laffont, J.J. & Tirole, J. (2000). *Competition in telecommunications*. Cambridge, MA: The MIT Press.

Levin, S.G., Levin, S.L., & Meisel, J.B. (1987). A dynamic analysis of the adoption of a new technology: the case of optical scanners. *Review of Economics and Statistics*, 69(1), 12-17.

Liebowitz, S.J. & Margolis, S.E. (1994). Network externality: An Uncommon Tragedy. *Journal of Economic Perspectives*, 8(2), 133-150.

Oren, S. & Smith, S. (1981). Critical mass and tariff structure in electronic communications markets. *Bell Journal of Economics*, 12(2), 467-486.

Rohlfs, J. (1974). A theory of interdependent demand for a communications service. *Bell Journal of Economics and Management Science*, 5(1), 16-37.

Rohlfs, J. (2001). *Bandwagon effects in high-technology industries*. Cambridge, MA: The MIT Press.

Taiwan Ministry of Transportation and Communications, The Directorate General of Telecommunications. (2000). *The Telecommunications Statistics Report*. Taipei, Taiwan.

Taiwan Ministry of Transportation and Communications, The Directorate General of Telecommunications. (2001). *The Telecommunications Statistics Report*. Taipei, Taiwan.

Valletti, T.M. (1999). A model of competition in mobile communications. *Information Economics and Policy*, 11(1), 61-72.

Part II. Decision Making with Respect to Diffusing Technologies

Chapter 4

WILLINGNESS-TO-PAY AND THE DEMAND FOR BROADBAND SERVICE*

Paul Rappoport
Temple University
Lester D. Taylor
University of Arizona
Donald J. Kridel
University of Missouri at St. Louis

I. INTRODUCTION

This study offers a variation on usual empirical studies of demand, in that price elasticities are derived directly from data on willingness-to-pay, rather than from coefficients estimated in an econometric demand function. The focus in the analysis is on the demand for broadband access to the telecommunications network, and makes use of data collected in a survey in which respondents were queried regarding the most that they would be willing-to-pay for cable-modem/DSL services. The framework used takes its cue from a largely overlooked procedure suggested by Cramer (1969), in which price elasticities are calculated from the cumulative distribution of willingness-to-pay. Price elasticities for both cable-modem and DSL access are obtained that range from an order of -3 for high access prices (and therefore low implied penetration rates) to less than 1 (in absolute value) for low access prices (and therefore high implied penetration).

These results are compared with our previous estimates of broadband elasticity. In general, the comparisons suggest that this procedure has merit and provides fertile ground for continuing research.

---

* The authors wish to thank Kenneth Fjell and Teodosio Perez-Amaral for helpful comments, Dale Kulp of Management Systems Group for access to the MSG omnibus survey, and Kevin Babyak of Econsult for research assistance.

II. THEORETICAL BACKGROUND

We begin with usual access/usage framework for determining the demand for access to a network, whereby the demand for access is determined by the size of the consumer surplus from usage of the network in relation to the price of access. Accordingly, let q denote usage, and let q(p,y) denote the demand for usage, conditional on a price of usage, p, and other variables (income, education, etc.), y. The consumer surplus (CS) from usage will then be given by

$$\text{Cs} = \int_p^\infty q(z,y)dz . \tag{1}$$

Let B denote the price of access. Access will then be demanded if

$$\text{CS} \geq \text{B}, \tag{2}$$

or equivalently (in logarithms) if

$$\ln\text{CS} \geq \ln\text{B}. \tag{3}$$

Perl (1983) was one of the first to apply this framework empirically, doing so by assuming a demand function of the form:

$$q = \text{A}e^{-\alpha p}y^{\beta}e^{u}, \tag{4}$$

where y denotes income (or other variables) and u is a random error term with distribution g(u). Consumer's surplus, CS, will then be given by

$$\text{CS} = \int_p^\infty Ae^{-\alpha z}y^{\beta}e^{u}dz = \frac{\text{A}e^{-\alpha p}y^{\beta}e^{u}}{\alpha} \tag{5}$$

With net benefits from usage and the price of access expressed in logarithms, the condition for demanding access to the telephone network accordingly becomes:

$$\text{P}(\ln\text{CS} \geq \ln\text{B}) = \text{P}(a - \alpha p + \beta\ln y + u \geq \ln \text{B}) = \text{P}(u \geq \ln\text{B} - a + \alpha p - \beta\ln y), \tag{6}$$

where $a = \ln(\text{A}/\alpha)$. The final step is to specify a probability law for consumer surplus, which, in view of the last line in equation (6), can be reduced to the specification for the distribution of u in the demand function for usage. An assumption that u is distributed normally leads to a standard probit model, while an assumption that u is logistic leads to a logit model. Empirical studies exemplifying both approaches abound in the literature.[1]

[1] Empirical studies employing the probit framework include Perl (1983) and Taylor and Kridel (1990), while studies using the logit framework include Bodnar *et al.* (1988) and Train, McFadden, and Ben-Akiva (1987). Most

The standard procedure for estimating access demand can thus be seen in terms of obtaining information on the consumer surplus from usage by estimating a demand function, and then integrating beneath this demand function. In the present context, however, our procedure is essentially the reverse, for what we have by way of information are statements on the part of respondents in a survey as to the *most* that they would be willing-to-pay for cable-modem or DSL service. This "*most*" accordingly represents (at least in principle) the maximum price at which the respondent would purchase that type of access. Thus, for any particular price of access, B*, say, access will be demanded for WTP's that are this value or greater, while access will not be demanded for WTP's that are less than this value. Hence, implicit in the *distribution* of WTP's is an aggregate demand function (or more specifically, *penetration function*) for cable-modem (or DSL) access. In particular, this function will be given by

$$\begin{aligned} D(B) &= \text{proportion of WTP's that are greater than or equal to } B = P(WTP \geq B) \\ &= 1 - CDF(B), \end{aligned} \tag{7}$$

where CDF(B) denotes the cumulative distribution function of the WTP's. Once CDF's of WTP's are constructed, price elasticities can be obtained (without intervention of the demand function) via the formula (or an empirical approximation thereof):

$$\text{Elasticity (B)} = \frac{d \ln CDF(\pi)}{d \ln \pi}. \tag{8}$$

III. Construction of CDF's for Cable-Modem and DSL Access

WTP CDF's for cable-modem and DSL access have been constructed from information collected in an omnibus survey of about 2000 households conducted in early 2002 by the Management Systems Group of Philadelphia. Each of the participants in the survey was asked one (but not both) of the following two questions regarding their willingness-to-pay for DSL and cable-modem access to the Internet:

(a). What is the most you would be willing to pay on a monthly basis for high-speed Internet service provided by your telephone/cable TV company?

(b). What is the highest monthly price at which you would consider purchasing high-speed Internet service provided by your telephone/cable TV company?

---

empirical studies of telecommunications access demand that employ a consumer-surplus framework focus on local usage, and accordingly ignore the net benefits arising from toll usage. Hausman, Tardiff, and Bellinfonte (1993) and Erikson, Kaserman, and Mayo (1995) are exceptions.

For the present analysis, CDF's are constructed under the assumption that responses elicited by the two questions are homogeneous with respect to the question, which is to say that "framing issues" (a la Kahneman-Tversky) are put to the side.[2]

A total of 2011 households participated in the MSG survey. However, elimination of various non-responses, together with refusals to provide information on income or other variables, reduced the number of usable responses to 1192. Exclusion of responses of "zero" for WTP then yields data sets of 975 observations for cable-modem and 922 observations for DSL that are used in constructing the CDF's for these quantities.[3] The resulting frequency distributions are given in Tables 1 and 2.

From the tables, we see that proffered WTP's ranged from highs of $400 and $390 per month for cable-modem and DSL, respectively, to a low of $5, with means, respectively, of $37 and $33 and standard deviations of $23 and $19.[4] Not surprisingly, "pileups" are seen to occur on WTP's divisible by 5 ($50, $45, $40, etc.). Moreover, in a number of cases, it is clear that the values elicited were simply the prices ($45.95, $39.99, $39.95, etc.) that respondents were currently paying for Internet access.[5]

IV. Calculation of Price Elasticities

As has been noted, an attractive feature of CDF's of willingness-to-pay is that they can be interpreted as penetration functions, from which it is possible to obtain price elasticities in line with expression (8) above. Three procedures come to mind as to how these elasticities might be constructed. The most straightforward procedure would be to define the elasticities as simple arc elasticities between selected adjacent values on the empirical CDF's. Unfortunately, however, because of the 'pileups' just noted, the elasticities that would emerge from this procedure would be highly unstable, and accordingly of little practical use. A second procedure would be to derive the elasticities on the presumption [a la Cramer (1969)] that the WTP's are distributed according to the log-normal distribution. This second procedure was our initial focus. The results obtained assuming that WTP's are distributed according to the log-normal distribution are reported in Section V. However, an analysis of the survey responses raised concerns that the assumption that the WTP's follow a log-normal distribution is empirically not supported by the data. With this concern in mind we adopted a less binding assumption to estimate elasticities by using a non-parametric approximation to the empirical CDF's. The results of this approach are presented in this Section.

---

[2] However, it is relevant to mention that, in a regression analysis in which the pooled responses to the two questions are related to income, age, education, and other factors, together with a dummy variable distinguishing the two questions, the dummy variable is of minor significance.

[3] The rationale for eliminating all "zero" responses for WTP is discussed below.

[4] WTP's of zero are, as noted, eliminated from the sample.

[5] It might be thought that the WTP's for cable-modem and DSL would be highly correlated, but this is not the case. A regression of the WTP for DSL (using the 923 DSL observations) on the WTP for cable-modem, with dummy variables for dial-up and cable-modem access, yields a coefficient of 0.50 for cable-modem WTP (with a t-ratio of 23) and an R2 of only 0.40. This is a curious result, for it suggests that cable-modem and DSL are not viewed by respondents as substitute forms of broadband access.

Table 1. Frequency Distribution for WTP's Cable-Modem Service

WTP	Freq.	%	Cum. Freq.	Cum. %	WTP	Freq.	%	Cum. Freq.	Cum. %
400	1	0.10	1	0.10	39.95	7	0.72	404	41.44
200	2	0.21	3	0.31	39	6	0.62	410	42.05
150	1	0.10	4	0.41	37	2	0.21	412	42.26
125	1	0.10	5	0.51	36	1	0.10	413	42.36
120	1	0.10	6	0.62	35	65	6.67	478	49.03
110	1	0.10	7	0.72	34.95	1	0.10	479	49.13
100	7	0.72	14	1.44	32.99	1	0.10	480	49.23
90	4	0.41	18	1.85	32	2	0.21	482	49.44
80	7	0.72	25	2.56	30	137	14.05	619	63.49
79	1	0.10	26	2.67	29.99	3	0.31	622	63.79
75	13	1.33	39	4.00	29.25	1	0.10	623	63.90
70	5	0.51	44	4.51	29	4	0.41	627	64.31
69	1	0.10	45	4.62	28	1	0.10	628	64.41
67	2	0.21	47	4.82	25	115	11.79	743	76.21
65	19	1.95	66	6.77	24.95	1	0.10	744	76.31
60	26	2.67	92	9.44	23.45	1	0.10	745	76.41
59.95	1	0.10	93	9.54	23	7	0.72	752	77.13
59	3	0.31	96	9.85	22	5	0.51	757	77.64
55	31	3.18	127	13.03	21.95	1	0.10	758	77.74
52	1	0.10	128	13.13	21	6	0.62	764	78.36
51	1	0.10	129	13.23	20.99	1	0.10	765	78.46
50	84	8.62	213	21.85	20	156	16.00	921	94.46
49.95	2	0.21	215	22.05	19.99	1	0.10	922	94.56
49	6	0.62	221	22.67	19.95	2	0.21	924	94.77
48	2	0.21	223	22.87	19.50	1	0.10	925	94.87
47	2	0.21	225	23.08	19	2	0.21	927	95.08
45	50	5.13	275	28.21	17	1	0.10	928	95.18
44.95	1	0.10	276	28.31	15	26	2.67	954	97.85
44	3	0.31	279	28.62	14.95	1	0.10	955	97.95
43	1	0.10	280	28.72	12.95	2	0.21	957	98.15
42	3	0.31	283	29.03	10	16	1.64	973	99.79
41	3	0.31	286	29.33	9.95	1	0.10	974	99.90
40	110	11.28	396	40.62	5	1	0.10	975	100.00
39.99	1	0.10	397	40.72					

Mean WTP: 36.86   Std. dev.: WTP: 21.82.

Table 2. Frequency Distribution for WTP's DSL Service

WTP	Freq.	%	Cum. Freq.	Cum. %	WTP	Freq.	%	Cum. Freq.	Cum. %
300	1	0.11	1	0.11	30	147	15.94	507	54.99
200	2	0.22	3	0.33	29.99	2	0.22	509	55.21
120	1	0.11	4	0.43	29.95	6	0.65	515	55.86
110	1	0.11	5	0.54	29	5	0.54	520	56.40
100	5	0.54	10	1.08	28	1	0.11	521	56.51
90	1	0.11	11	1.19	26	2	0.22	523	56.72
80	5	0.54	16	1.74	25	117	12.69	640	69.41
75	3	0.33	19	2.06	24.99	1	0.11	641	69.52
70	2	0.22	21	2.28	23	10	1.08	651	70.61
69	2	0.22	23	2.49	22	6	0.65	657	71.26
67	1	0.11	24	2.60	21.99	1	0.11	658	71.37
65	9	0.98	33	3.58	21.95	2	0.22	660	71.58
60	15	1.63	48	5.21	21	6	0.65	666	72.23
59	1	0.11	49	5.31	20	173	18.76	839	91.00
57	1	0.11	50	5.42	19.99	3	0.33	842	91.32
55	21	2.28	71	7.70	19.95	4	0.43	846	91.76
50	75	8.13	146	15.84	19	4	0.43	850	92.19
49	3	0.33	149	16.16	18	1	0.11	851	92.30
48	2	0.22	151	16.38	17	1	0.11	852	92.41
47	1	0.11	152	16.49	16	2	0.22	854	92.62
45.95	1	0.11	153	16.59	15	27	2.93	881	95.55
45	31	3.36	184	19.96	14.95	2	0.22	883	95.77
43	1	0.11	185	20.07	14	2	0.22	885	95.99
41	2	0.22	187	20.28	12.95	2	0.22	887	96.20
40	88	9.54	275	29.83	12	1	0.11	888	96.31
39.99	1	0.11	276	29.93	10	29	3.15	917	99.46
39.95	4	0.43	280	30.37	9.95	1	0.11	918	99.57
39	14	1.52	294	31.89	7	1	0.11	919	99.67
38	2	0.22	296	32.10	6	1	0.11	920	99.78
36	2	0.22	298	32.32	5	2	0.22	922	100.00
35	54	5.86	352	38.18					
34	4	0.43	356	38.61					
32	3	0.33	360	39.05					
			Mean WTP: 32.93			Std. dev.: 18.89.			

Specifically, what we have done is to estimate kernel density functions (using the unit normal density function as the kernel weighting factor) as approximations to the empirical frequency distributions of the WTP's, and then calculating elasticities at various points on these functions. Since the use of kernel estimation may be seen as somewhat novel in this context, some background and motivation may be useful. The goal in kernel estimation is to develop a continuous approximation to an empirical frequency distribution that, among other things, can be used to assign density, in a statistically valid manner, in any small neighborhood of an observed frequency point. For example, in Table 1, we see that there are 86 observations at a WTP of \$50 (treating \$49.95 as \$50), but only a single observation at \$51 and six at \$49. Such "lumpiness", as was noted in the preceding paragraph, makes calculation of an arc elasticity in a neighborhood of \$50 a dubious exercise.

Since there is little reason to think that, in a large population, "pileups" of WTP's at amounts divisible by \$5 reflect anything other than the convenience of nice round numbers, there is also little reason to think that the "true" density at WTP's of \$51 or \$49 ought to be much different than the density at \$50. The intuitive way of dealing with this contingency (i.e., "pileups" at particular discrete points) is to tabulate frequencies within intervals, and then to calculate "density" as frequency within an interval divided by the length of the interval (i.e., as averages within intervals). However, in doing this, the "density" within any particular interval is calculated using only the observations within that interval, which is to say that if an interval in question (say) is from \$40 to \$45, then a WTP of \$46 (which is as "close" to \$45 as is \$44) will not be given weight in calculating the density for that interval. What kernel density estimation does is to allow *every* observation to have weight in the calculation of the density for *every* interval, but a weight that varies inversely with the "distance" that the observations lies from the center of the interval in question.

For the analytics involved, let $\hat{g}(x)$ denote the density function that is to be constructed for a random variable x (in our case, WTP) that varies from x_1 to x_n. For cable-modem WTP, for example, the range x_1 to x_n would be \$5 to \$400. Next, divide this range (called the 'support' in kernel estimation terminology) into k sub-intervals. The function $\hat{g}(x)$ is then constructed as:

$$\hat{g}(x_i) = \sum_{j=1}^{N} \frac{K\left(\dfrac{x_i - x_j}{h}\right)}{Nh}, \quad i = 1, \ldots, k. \tag{9}$$

In this expression, K denotes the kernel-weighting function, h represents a smoothing parameter, and N denotes the number of observations. For the two cases at hand (for cable-modem and DSL WTP's), the density function in expression (9) has been constructed for each using the unit normal density function as the kernel weighting function and a 'support' of k = 1000 intervals.[6]

With estimation of the kernel density functions for cable-modem and DSL WTP's, price elasticities can be calculated from the associated CDF's [using a numerical counterpart to expression (8)]. The resulting elasticities, calculated at prices (WTP's) of \$70, \$60, \$50, \$40, \$30, and \$20, are presented in Table 3.[7]

---

[6] Silverman's rule-of-thumb,

$$\hat{h} = (0.9)\min[\text{std. dev., interquartile range}/1.34](N\text{-}1/5),$$

has been used for the smoothing parameter h. Two standard references for kernel density estimation are Silverman (1986) and Wand and Jones (1995). Ker and Goodwin (2000) provide an interesting practical application to the estimation of crop insurance rates.

[7] The values for the cumulative densities at \$70, \$60, etc. are taken from the CDF's for the empirical frequency distributions in Tables 1 and 2. The elasticities are then calculated from corresponding points on the kernel

From the table, we see that the elasticities range from less than -3 to a little greater than -1. However, since (from Tables 1 and 2) half of penetration occurs for WTP of \$30 or less for both forms of broadband access, the most relevant elasticity can be taken to lie in the range of -2 to -1. As will be seen below, this range compares with existing econometric estimates that vary from -0.84 to -1.05.[8]

Table 3. Price Elasticities for Broadband Access Cable-Modem and DSL

	Elasticity	
WTP	**Cable-Modem**	**DSL**
\$70	-2.97	-2.21
60	-3.29	-2.57
50	-3.17	-2.99
40	-2.35	-3.10
30	-1.53	-2.10
20	-0.95	-0.97

In an earlier version of this study, given at the ICFC Conference in San Francisco in June 2002 (also at the European Conference of the International Telecommunications Society in Madrid in September 2002), elasticities were calculated assuming that the WTP's are distributed log-normally [as per Cramer (1969)]. Since the empirical frequency distributions provide little support for the assumption that the WTP's are log-normal, our preference is to base elasticities on a kernel-smoothed density estimator as just discussed. However, for illustrative purposes, comparable elasticities will now be derived for cable-modem access under a log-normal assumption.

From the last line in expression (7) above, we will have for a log-normal demand function:

$$
\begin{aligned}
D(B) &= 1 - CDF(B) \\
&\ 1 - \Lambda(B; \mu, \sigma^2),
\end{aligned}
\qquad (9)
$$

CDF's via the formula (where x denotes the smallest value on the kernel CDF that is greater than x, for x = \$70, ...,\$20.):

$$Elasticity(x) = \frac{\Delta\, CDF(x) \,/\, CDF(x)}{\Delta\, WTP(x) \,/\, WTP(x)}$$

[8] See, e.g., Rappoport, Taylor, Kridel, and Serad (1998), Kridel, Rappoport, and Taylor, (1999), Kridel, Rappoport, and Taylor (2001), Rappoport, Kridel, Taylor, Duffy-Deno, and Alleman (2002), and Rappoport, Taylor, and Kridel, (2002). In the earlier version of this study, given at the ICFC Conference in San Francisco in June 2002 (also at the European Conference of the International Telecommunications Society in Madrid in September 2002), elasticities were calculated assuming that the WTP's are distributed log-normally [as per Cramer (1969)]. The elasticities obtained by this procedure are, for WTP's of \$50 or less, about half those obtained here. However, the empirical frequency distributions provide little support for the assumption that the WTP's are log-normal, hence our preference to use a kernel-based density estimator. Also, the earlier data set included the responses of zero WTP's, which have been excluded in the present data set. Much of the difference is almost certainly due to this exclusion. The decision to exclude zeros was based on the consideration that a response of zero WTP was an expression that the value of cable-modem or DSL access was zero even if access were free, rather than that access would be demanded by everyone at prices of zero.

where $\Lambda(B; \mu, \sigma^2)$ denotes the CDF for the log-normal distribution for WTP, with mean equal to μ [= mean(lnWTP)] and variance equal to σ^2 [= var(lnWTP)]. The elasticity at any price B on this function will then be given [from expression (8)] by:

$$\text{Elasticity (B)} = -\frac{1}{\sigma}\frac{n(v;0,1)}{N(v;0,1)}, \tag{10}$$

where n(v;0,1) and N(v;0,1) denote the density and CDF functions for the unit normal distribution, and

$$v = \frac{\ln \pi - \mu}{\sigma}. \tag{11}$$

Log-normal elasticities, as derived from expression 10, using the survey mean and variance of lnWTP as estimates of μ and σ^2, are presented in Table 4 at the same price points as in Table 3. Comparing Tables 3 and 4, we see that the log-normal elasticities are similar to their kernel-based counterparts at prices of \$60 and \$70, but are markedly smaller at prices of \$50 and less. A likely reason for the smaller log-normal values for \$50 and less is the log-normal calculations use an alternatively edited data set that includes inclusion of the zero values for WTP in the log-normal calculations.[9]

Table 4. Price Elasticities for Cable-Modem Access

Price	Elasticity
\$20	-0.53
\$30	-0.59
\$40	-0.75
\$50	-0.98
\$60	-2.25
\$70	-3.34

As noted in footnote 9, the present analysis is simply the latest exercise in a continuing study of Internet demand. Table 5 provides a summary of our previous elasticity estimates regarding Internet access. Ranges are provided where there were alternative specifications and models. The values reported in this paper are generally comparable.

[9] The inclusion/non-inclusion of the zero values for WTP in the calculations is a debatable point (even amongst ourselves). If, as was the case in the San Francisco ICFC presentation, the focus is on the derivation of a market demand function, the inclusion of zero values is clearly legitimate, since these provide an upper bound to overall market penetration. However, if, as in the present exercise, the focus is more on the estimation of price elasticities (for that part of the market with a non-zero demand), then it seems that zero WTPs ought to be excluded.

Table 5. Previous Estimated Internet Elasticities

Paper	Dial-up Access	Cable-modem	DSL	Broadband
Rappoport, Taylor, Kridel (2002a)				-1.50
Rappoport, Kridel, Taylor, Duffy-Deno, Alleman (2002)	-.17 to -.28	-1.35	-1.79	-1.49
Kridel, Rappoport, Taylor (2001)		-1.08 to –1.79		
Kridel, Rappoport, Taylor (1999)	-.18 to -.38			

V. Conclusions

In this paper, we have utilized information on willingness-to-pay for cable-modem and DSL access to the telecommunications network obtained from an omnibus survey of more than 2000 households to construct estimates of price elasticities for broadband access to the Internet. The elasticities obtained are generally similar for the two forms of access, and range from roughly -1.00 at a monthly access price of $20 to -3.00 or less for access prices of $40 or more. Since these elasticities are constructed from information elicited directly from households, and thus entail the use of contingent-valuation (CV) data, the seriousness (in light of the longstanding controversy surrounding the use of such data) with which our elasticities are be taken might be open to question.[10] However, in our view, the values that we have obtained are indeed plausible and warrant serious consideration. This is especially the case for the values calculated in the part of the WTP CDF's for which half or more of broadband penetration would be projected to occur (that is, at access prices of $30 or less, as per Tables 1 and 2), where the CV-based elasticities are not notably out-of-line with existing econometric estimates. Added credence for our results, it seems to us, is provided by the fact that, with cable-modem and DSL, we are dealing with products (or services) with which respondents are already familiar (and in many instances already demand), unlike in circumstances (such as in the valuation of a unique natural resource or the absence of a horrendous contingent event) in which there is no generally meaningful way that valuation can be assessed. Finally, these results suggest that the pricing of broadband access is still a major issue with respect to broadband adoption. Significant increases in broadband penetration rates will mostly likely require large price reductions for access. The future pricing of content is another matter for future research.

[10] The critical literature on contingent valuation methods is large. See the NOAA Panel Report (1993), Smith (1993), Portnoy (1994), Hanneman (1994), Diamond and Hausman (1994), and McFadden (1994). On the other hand, particularly successful uses of CV data would seem to include Hammitt (1986) and Kridel (1988). In the present context, one should be skeptical of the largest WTP's, especially for cable-modem access. The respondent providing the highest value ($400) has cable-modem access and an income between $75,000 and $100,000, and thus would not appear to be problematic. However, the next four highest cable-modem WTP's did not report any form of Internet access, and one of them (with a WTP of $200) reported income of less than $15,000. Of course, the latter (who reported an age of 27) may have had high-speed access at work or school or at a friend's or parents' house, and was reporting what his/her WTP for cable-modem access would be in the absence of any budget constraint.

References

Bodnar, J., Dilworth, P., and Iacono, S. (1988), "Cross-Section Analysis of Residential Telephone Subscription in Canada," *Information Economics and Policy*, Vol. 3, No. 4, pp. 311-331.

Cramer, J.S. (1969), *Empirical Econometrics*, Elsevier Publishing Co., New York.

Diamond, P.A. and Hausman, J.A. (1994), "Contingent Valuation: Is Some Number Better Than No Number?," *Journal of Economic Perspectives*, Volume 8, No. 4, Fall 1994, pp. 45-64.

Erikson, R.C., and Kaserman, D.L., and Mayo, J.W. (1998), "Targeted and Untargeted Subsidy Schemes: Evidence from Post-Divestiture Efforts to Promote Universal Service," *Journal of Law and Economics*, Vol. 41, October 1998, pp. 477-502.

Hammitt, J.K. (1986), "Estimating Consumer Willingness to Pay to Reduce Food Borne Risk," *Report R-3447-EPA*, The RAND Corporation.

Hannemann, W.M. (1994), "Valuing the Environment though Contingent Valuation," *Journal of Economic Perspectives*, Volume 8, No. 4, Fall 1994, pp. 19-44.

Hausman, J.A., Tardiff, T.J., and Bellinfonte, A. (1993), "The Effects of the Breakup of AT&T on Telephone Penetration in the United States," *American Economic Review Papers*, Vol. 83, No. 2, May 1993, pp. 178-184.

Ker, A.P. and Goodwin, B.K. (2000), "Nonparametric Estimation of Crop Insurance and Rates Revisited," *American Journal of Agricultural Economics*, Vol. 83, May 2000, pp. 463-478.

Kridel, D.J. (1988), "A Consumer Surplus Approach to Predicting Extended Area Service (EAS) Development and Stimulation Rates," *Information Economics and Policy*, Vol. 3, No. 4, pp. 379-390.

Kridel, D.J., Rappoport, P.N., and Taylor, L.D. (1999), "An Econometric Model of the Demand for Access to the Internet," in *The Future of the Telecommunications Industry: Forecasting and Demand Analysis*, ed. by D.G. Loomis and L.D. Taylor, Kluwer Academic Publishers.

Kridel, D.J., Rappoport, P.N., and Taylor, L.D. (2001), "An Econometric Model of the Demand for Access to the Internet by Cable Modem," *Forecasting the Internet: Understanding the Explosive Growth of Data Communications*, ed. by D.G. Loomis and L.D. Taylor, Kluwer Academic Publishers.

National Oceanographic and Atmospheric Administration (NOAA;1993), 58, *Federal Register*, 4601, January 15, 1993.

Maddala, G.S. (1969), *Limited –Dependent and Qualitative Variables in Econometrics,* Cambridge University Press.

McFadden, D. (1994), "Contingent Valuation and Social Choice," *American Journal of Agricultural Economics*, Vol. 76, November 1994, pp. 695-707.

Portnoy, P.R. (1994), "The Contingent Valuation Debate: Why Economists Should Care," *Journal of Economic Perspectives*, Volume 8, No. 4, Fall 1994, pp. 3-18.

Rappoport, P.N., Taylor, L.D., and Kridel, D.J. (2002a), "The Demand for Broadband: Access, Content, and the Value of Time," in *Broadband: Should We Regulated High Speed Internet Access*, ed. by J. Alleman and R. Crandall, Brookings, 2002.

Rappoport, P.N., Kridel, D.J., Taylor, Duffy-Deno, K., and Alleman, J. (2002), "Forecasting The Demand for Internet Services," in *The International Handbook of*

Telecommunications Economics: Volume II, ed. by G. Madden, Edward Elgar Publishing Co., London.

Rappoport, P.N., Taylor, L.D., Kridel, D.J., and Serad, W. (1998), "The Demand for Internet and On-Line Access," in *Telecommunications Transformation: Technology, Strategy and Policy*, ed. by E. Bohlin and S.L. Levin, IOS Press.

Smith, V.K. (1993), "Non-Market Valuation of Natural Resources: An Interpretive Appraisal," *Land Economics*, Vol. 69, No. 1, February 1993, pp. 1-26.

Taylor, L.D. and Kridel, D.J. (1990), "Residential Demand for Access to the Telephone Network," in *Telecommunications Demand Modeling*, ed. by A. de Fontenay, M.H. Shugard, and D.S. Sibley, North Holland Publishing Co., Amsterdam.

Train, K.E., McFadden, D.L, and Ben-Akiva, M. (1987), "The Demand for Local Telephone Service: A Fully Discrete Model of Residential Calling Patterns and Service Choices," *The Rand Journal of Economics*, Vol. 18, No. 1, Spring 1987, pp. 109-123.

Chapter 5

TECHNOLOGICAL CHANGE AND FIRST-CLASS LETTER MAIL*

Peter Bernstein
RCF Consulting Inc.

I. INTRODUCTION AND OVERVIEW

Technological change has had both positive and negative impacts on the United States Postal Service. During the 1970s and 1980s, the spread of business computers and data base management tools probably increased mail volume, as it made direct mail advertising less costly and more effective. Other developments, such as the introduction of cable-TV and cell phones, along with deregulation of the financial system, increased the number of bills, payments, and statements traveling through the mail system.

But recent technological changes appear to be having a net negative impact on mail volume. These changes, many of which are related to the spread of the Internet, are having their greatest effects on the Postal Service's largest revenue source, First-Class letter mail. First-Class letter mail, which consists primarily of correspondence, transaction, and advertising mailings, is particularly vulnerable to competition from new technologies. E-mail has become a common alternative to communication by mail, bills can be paid online or through an automatic deduction from a bank account, Internet advertising is a substitute from direct mail.

The substitution of technological alternatives for the mail is referred to as "technological diversion" or "electronic diversion." This chapter examines the diversion of letter mail, presents estimates of the magnitude of past diversion, details work done to model this diversion, and provides both near-term and longer-term forecasts of the future diversion of letter mail. Much of this chapter is taken from testimony presented on behalf of the Postal Service before the Postal Rate Commission in the R2001-1 rate case, filed in September of 2001. As a result, some of the data and analyses presented here are already obsolete,

* Several people helped me with the preparation of the original testimony and provided useful comments on earlier drafts of this chapter. They include George Tolley and Thomas Thress of RCF, Inc. and Eric Koetting, Grady Foster, and Malcolm Harris of the Postal Service.

especially considering the dramatic events that affected the country, the economy, and the Postal Service following that rate case filing. Nonetheless, most of the analysis is still relevant, as it reveals how technological developments affect another, competing entity, and how that entity assesses and responds to that challenge.

The outline of this chapter is as follows. Following this introduction, section II provides the reader with an overview of postal finances and the history of volumes of major classes of mail. Section II also includes a brief discussion of the postal rate-making process and with that, a review of past attempts to model "electronic diversion" of mail volumes. Section III discusses some of the key technological drivers and presents some concrete examples of how the mail has been affected by the use of alternatives. Section IV describes work done to project future diversion through development of models to predict future use of Internet-related technologies. In Section V, estimates of past and present diversion of letter mail are presented, along with projections of future losses, both in the near term and the long-term. This chapter also discussed some of the Postal Service's responses to increased competition. A final section presents a brief look at relevant recent events and provides some concluding thoughts.

II. Understanding Postal Finances

In fiscal 2001, the United States Postal Service had revenues of nearly $70 billion, greater than Microsoft and AOL-Time Warner combined.[1] Mail volume in 2001 surpassed 200 billion pieces, well more than two and a half times the volume in 1970, when the Postal Reorganization Act created the Postal Service from what had been the Post Office department of the federal government.

A. The Act of 1970 and Afterwards

The Postal Reorganization Act of 1970 created the Postal Service as a quasi-independent agency within the federal government. The 1970 Act mandated that the Postal Service be run in a "business-like" manner, required that the agency be self-supporting free of government subsidies, and established the Postal Rate Commission to serve as a regulatory authority over postal rates.

In the 30 plus years since the 1970 Act, the total volume of mail delivered by the Postal Service increased from 80.7 billion pieces in 1970 to 207.2 billion pieces in 2001, an average annual growth rate of 3.1 percent. Postal revenues grew from $5.9 billion to $65.6 billion, an annual average increase of more than eight percent.. Government subsidies, which represented nearly 20 percent of postal revenues in 1970, have vanished. Postal revenues have been approximately equal to costs, with the agency reporting a positive net income as recently

[1] A peculiarity of the Postal Service is that it operates under two different fiscal years. The 2001 government fiscal year (GFY 2001) began October 1, 2000 and ended September 30, 2001. The postal fiscal year is a 52 week (364 day) year that begins and ends on different days in different years. PFY 2001 began September 9, 2000 and ended September 7, 2001. Unless otherwise noted, all fiscal year references are to postal fiscal years. The difference between PFYs and GFYs is rarely important.

as 1999. Postal rates, which increased sharply in the early 1970s as government subsidies were reduced, have remained essentially constant in real terms since 1976.

Table 1 shows Postal Service revenues as a share of GDP from 1970 to 2001. Over this period, the postal revenue share has ranged from 0.57 percent to 0.73 percent. Revenue share grew in the 1980s and early 1990s, but has been declining since 1996. Table 1 also shows that most of the decline in the postal revenue share is due to declines in the revenue share from First-Class letters, which fell from 0.43 percent of total GDP in 1996 to 0.35 percent of GDP in 2001.

Table 1. Postal Service Revenues as Share of GDP

Year	Postal Revenues as Share of GDP	First-Class Revenues as Share of GDP	All Other Revenues as Share of GDP
1970	0.57%	0.34%	0.23%
1975	0.60%	0.37%	0.23%
1980	0.59%	0.37%	0.22%
1985	0.64%	0.40%	0.24%
1990	0.66%	0.41%	0.25%
1995	0.73%	0.43%	0.30%
1996	0.73%	0.43%	0.30%
1997	0.70%	0.41%	0.29%
1998	0.68%	0.39%	0.29%
1999	0.68%	0.38%	0.30%
2000	0.65%	0.36%	0.29%
2001	0.64%	0.35%	0.29%

Table 2 reveals the main reason for this decline, a slowdown in the growth rate of mail volume. Whereas from 1980 to 1990, volume grew at an annual average rate of 5.1 percent, growth declined to an average of just 2.2 percent per year from 1990 to 2000. This volume slowdown has occurred across most mail classes, but is particularly evident with First-Class Mail, for which average annual volume growth fell from 4.1 percent to 1.4 percent across the two decades. This trend continued into 2001, with volume growing less than one percent. Table 2 also shows the rapid growth of Standard Mail in the 1980s, a decade during which its volume doubled, in part as a result of technological changes benefiting the mail.

Table 2. Average Annual Growth in Mail Volume

	1980 to 1990	1990 to 2000
Total Mail	5.1%	2.2%
First-Class Mail	4.1%	1.4%
Periodicals	0.6%	-0.4%
Standard Mail	7.9%	3.5%
Packages	3.3%	6.8%

B. First-Class Letters

First-Class letters is the "bread and butter," of the Postal Service.[2] In Fiscal 2001, it accounted for half of all mail volume and somewhat more than half of all postal revenue, as shown in Table 3. The prominence of First-Class letters is in part due to the special protections it receives as dictated by the Private Express Statutes. The Private Express Statutes essentially prohibit the private delivery of letter mail. The statutes are generally seen as being a function of the Postal Service's obligation to provide universal service at a uniform price. It is argued that in the absence of these protections, private competitors could engage in a fairly easy form of "cream-skimming," providing service to only the lower cost markets at a price below the uniform letter rate, thereby leaving the Postal Service with only the higher cost delivery operations.

Table 3. 2001 Postal Volume and Revenue by Class of Mail (In billions)

	2001 Volume	Share of Total	2001 Revenue	Share of Total
First-Class Mail	103.8	50.1%	$35.9	54.7%
Periodicals Mail	10.1	4.9%	$2.2	3.4%
Standard Mail	90.6	43.7%	$15.7	23.9%
Packages	2.3	1.1%	$7.9	12.0%
Other	9.5	4.6%	$3.9	5.9%
Total	207.2	100.0%	$65.6	100.0%

While letter mail is protected from private delivery, it is not protected from other, less direct, forms of competition. In the past, the telephone operated as a substitute for letter mail. Now, letter mail substitutes include the many technological alternatives to be addressed in this chapter including, for example, fax transmissions, e-mail messages, and online bill payment. That these technological alternatives have adversely affect First-Class letter volume growth is evident from the data presented in Table 2.

C. Postal Rate Making

Postal rate making is a complicated process and this section provides only a general overview of the many issues involved. The Act of 1970 constituted the Postal Service as a break-even enterprise, with revenues from operations offsetting costs. When the Postal Service finds that projected future revenues are insufficient to cover projected future costs, it files for a rate increase before the Postal Rate Commission (PRC). The filing begins a ten-month process during which the PRC reviews testimony submitted by the Postal Service and intervening parties, before making a rate recommendation to the Postal Board of Governors. In most cases, the Board approves the recommended rates to go into effect usually within two months, or about one year after the case filing. In some cases, the Board of Governors may

[2] First-Class letters is actually a subclass within First-Class Mail, which also includes First-class cards (postcards) which account for about one-twentieth of First-Class volumes and revenues.

approve only part of the PRC recommendation and send back other parts for further analysis of yet unresolved disputes.

Oftentimes the cost, volume, or revenue projections of the Postal Service are challenged by intervening parties. Beyond questions related to the overall financial projections, intervening parties also lobby to shift the burden of any required rate increase from one class of mail to another. In that sense, much of the process is essentially a zero-sum game, with different parties proposing different ways to close the projected deficit.

The PRC makes a final rate recommendation taking into consideration all information presented in the case. Postal rates are subject to nine rate-making criteria, only one of which is binding – the prohibition against cross-subsidization. The rate for each postal product must cover the costs associated with that product, though in practice there are times when product revenues fail to cover product costs in a given year.

Postal costs are categorized as being volume variable or non-volume variable. The vast majority of non-volume variable costs are viewed as being common, or fixed, costs. Since total costs exceed volume-variable costs, prices must be set above volume-variable cost per piece, which serves as a proxy for marginal cost. The mark-up above marginal cost is determined by a number of factors including elasticity considerations, the current mark-up, various interpretations of the rate-making criteria and the relative success of the various intervening parties.

Table 4 shows the prices, unit volume variable costs, and percentage mark-up for different mail products in 2001. Prices and unit costs are often averages of a number of individual mail products contained within a given class of mail.

Table 4. 2001 Postal Prices, Unit Costs, and Percentage Mark-ups

	Average Price	Average Unit Cost	Percentage Mark-up
First-Class Mail	$0.346	$0.171	102.0%
Periodicals Mail	$0.219	$0.226	-2.3%
Standard Mail	$0.175	$0.111	57.2%
Packages	$3.437	$2.407	42.8%
Total	$0.317	$0.185	71.1%

First-Class letters has a relatively high mark-up. This is in part due to its relatively low demand elasticity, part due to the view that letter mail is responsible for the universal service obligation and the common delivery network, and part due to the economic realities of the break-even requirement. One consequence of the higher mark-up on First-Class letters is that the net income or loss of the Postal Service is driven by changes in letter volume, more so that by changes in the volume of other products. This is one reason why the diversion of First-Class volume to technological alternatives carries such importance to the Postal Service.

D. Mail Volume Forecasting

Of the myriad testimonies and documents submitted by the Postal Service in a rate filing, some of the most important ones are the testimonies related to projections of future mail volumes. For the past twenty years, mail volume forecasts have been based on econometric analyses performed by Dr. George Tolley and his colleagues at RCF Economics in Chicago. In addition to being used in postal rate cases, the forecasts are also used for internal purposed by Postal Service management. In general, mail volume forecasts have been quite accurate, relying primarily on econometric estimates of the elasticity of volume to changes in price, income, population, and other factors including various postal rule changes that have occurred over the years. However, forecasts of First-Class letter volume for 2001, presented in the R2000-1[3] rate case and in Table 5 below, were found to be exceed actual volumes by an appreciable amount. Furthermore, this over-forecast was entirely in the volume of what is known as single-piece letters, the basic one ounce or less letter product.

Table 5. Forecasted and Actual Letter Volumes (in billions) for 2001

	Forecasted Volume	Actual Volume	Piece Difference	Percent Difference
Total First-Class Letters	99.858	98.262	-1.596	-1.6%
Single-Piece	52.878	51.253	-1.625	-3.2%
Workshare	46.980	47.009	+0.029	0.1%

At that time, the postal forecasting equations did not include an explicit measure of the impact of new technologies on mail volume. Time trends were included in the analysis, but these terms were designed to measure longer-term developments such as the gradual decline in the use of personal letters and greeting cards. The volume changes noted above appeared to be due to more recent developments related to the greater use of technological alternatives to letter mail. This observation led to the development of new testimony for the Postal Service's R2001-1 rate case, focusing on the diversion of mail, and from which much of the remainder of this chapter is taken.

III. TECHNOLOGY REVIEW

A. Examples of Technological Diversion

Technological (or electronic) diversion is the term given to the loss of mail volume to technological alternatives. Table 6 shows a decomposition of First-Class letter mail, by type of mail and by sender and recipient. HH refers to households, NHH refers to non-households (business and government). Also shown is a list of some of the new technological alternatives to each type of First-Class letter mail. Some of these sources of letter diversion are fairly long-standing such as electronic funds transfers (EFT) and electronic data interchange (EDI). Others are fairly new and may have only a small impact on mail volumes at present (e.g., electronic bill presentment and payment or EBPP). Still, virtually every type of letter mail is subject to diversion to electronic alternatives.

[3] Postal rate cases are referenced by the year in which they are filed. Usually, volume forecasts are made for the following year, based on data from the previous year. For example, the R2000-1 rate case included forecasts of 2001 volumes, based on 1999 (and earlier) actual data.

Table 6. Examples of Technological Alternatives to First-Class Letter Mail

Type of Letter Mail	Share of Letter Volume	Recent Technological Alternatives
HH to HH		
Correspondence	7.0%	E-Mail
NHH to HH		
Bills and Statements	22.5%	Online Banking and EBPP
Advertising	7.5%	Internet Advertising
Other	15.0%	E-Mail and Fax
HH to NHH		
Bill Payments	11.0%	Computer Bill Pay, EFT
Other	2.0%	E-Mail and Fax
NHH to NHH		
Bills and Statements	20.0%	EDI, Online Banking, EBPP
Advertising	5.0%	Internet Advertising
Other	10.0%	E-Mail and Fax

B. Drivers of Technological Diversion

There are many drivers of technological diversion, and the table above shows that not all of these are directly related to the expansion of the Internet. Still, the Internet is uniquely important. Prior to the Internet, there were basically two ways to transmit individual information across the country, the telephone and the mail. The Internet has become a third national communication network.

In addition, Internet access is subject to network effects, sometimes called network externalities. This is the feature that as the number of participants in a network increases, the number of possible connections increases by almost the square of the increase in network participation. As more people and businesses have online connections, the importance of the Internet grows, and more and more opportunities become available online. It is now taken for granted in many circles that individuals have E-mail accounts, perhaps even more than one. Correspondence by E-mail is often assumed to be possible today, whereas in the past it was not. In fact, for the technologically more experienced, the word "mail" refers to E-mail, not postal mail. The spread of E-mail might have the result of opening the door to other forms of non-mail communication. Thus, many of the drivers of technological diversion discussed below are linked, either directly or indirectly, to the Internet.

1. Household Computer Ownership

Household computer ownership is one of the important drivers of technological diversion. Table 7 shows the growth in the percentage of households with a personal computer and a modem from 1990 through 2001 using data from the Postal Service's annual surveys of between 5,000 and 10,000 households. Table 7 shows that more than half of all households had this technology in 2001, nearly twice the level just three years ago. Moreover, the computers of today are faster and more powerful than in the past, allowing for a wider range of activities. Though much of this development has little to do with traditional uses of the mail (e.g. high-definition graphics for computer-gaming), high powered computers are a key input into the use of many alternatives to the mail.

Table 7. Percentage of Households with Personal Computer and Modem

1990	1991	1992	1993	1994	1995	1996	1997	1998	1999	2000	2001
5.2	6.4	7.2	8.9	11.2	14.4	19.1	22.1	28.8	35.3	46.5	55.5

Source: United States Postal Service, Household Diary Studies (1990 – 2001)[4]

2. Online Population

A second key driver of technological competition is the online population. According to Nielsen/NetRatings, as of July 2001, 165.2 million Americans had at-home Internet access, or approximately 58 percent of the population.[5] The article reports that "Fifty eight percent of all Americans had Internet access in their homes in July 2001, compared to 52 percent last year. In July 1999 only 39 percent of all Americans (106.4 million people) had access to the Web."

The Nielsen numbers for 2001 represent a tremendous increase in the online population over the last few years. Jupiter Media Metrix reports that just 37 million Americans were online in 1996 and 83 million online in 1998, meaning that the online population has doubled in the past three years, and more than quadrupled from 1996 to 2001.

Still, many people with access rarely use the Internet. A more telling figure than the online population (defined as those with access to the Internet) is the number of active users, defined by Nielsen as those who have used the Internet at home over the past 30 days. Chart A shows monthly active Internet users from April of 1999 through May of 2001, during which time active users at home increased from 57.6 million to 103 million.

A related measure is the number of households subscribing to Internet services. According to a census from Telecommunications Reports International (TRI), there were 70.7 million household subscribers to Internet services in the second quarter of 2001. This compares with 50.3 million subscribers in the first quarter of 2000, which was itself an 11 percent increase from the fourth quarter of 1999.[6]

In addition to the number of people online, the length of Internet experience is also important. Typically, the longer that a person has a computer and Internet access, the greater his or her range of uses. Household Diary Study data reveals that one in six Internet users have at least five years experience and two-thirds of users have at least two years experience. A related metric is time spent online. According to Nielsen, time spent per active user at home increased from seven hours per month in April of 1999 to more than 10 hours per month in July of 2001.

[4] The Diary Study provides information on household use of the mail as well as a variety of data on household demographics.

[5] US Internet Audience Up 16 Percent in Past Year, CyberAtlas, August 13, 2001.

[6] CyberAtlas, "Number of US Households Online Grows in Second Quarter," August 8, 2001 and CyberAtlas, "5 Million US Customers Go Online in Q1," May 2, 2000.

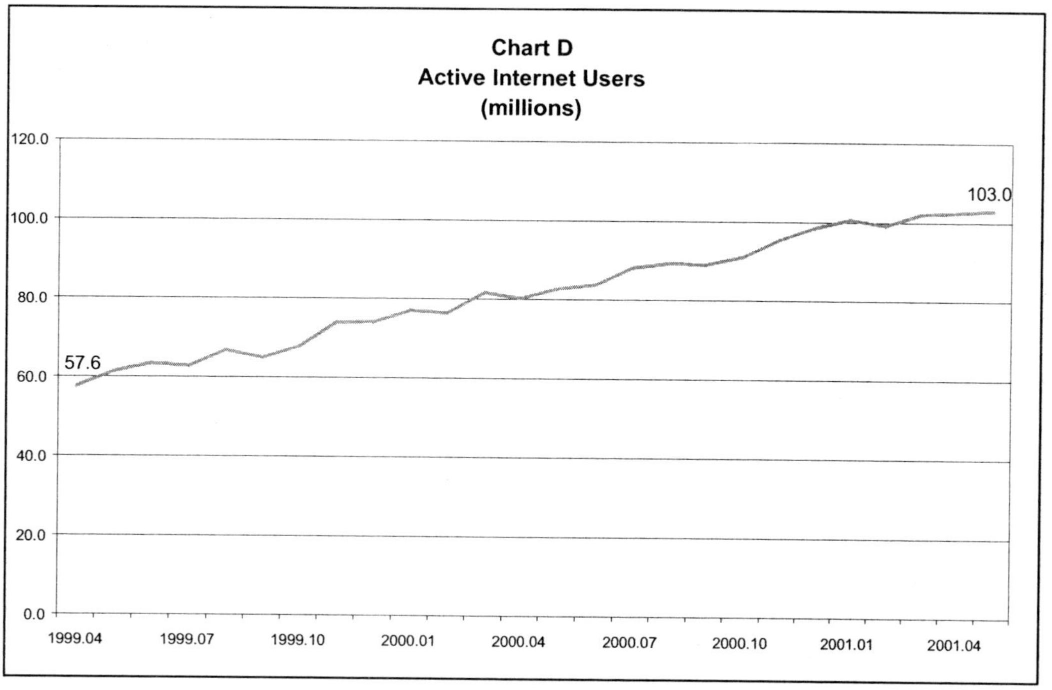

3. Broadband Technology

A third driver of technological diversion is the spread of broadband technology. Broadband technology allows access to the Internet that is much faster than through traditional dial-up modems using either cable wires or DSL (digital subscriber lines) connections. Not only is broadband faster, but it is also gives the user constant connection without dial-up waiting and without occupying a phone line. The increased speed has many advantages, one of which is faster display of graphical content such as scanned bills from an EBPP service. It also allows for what is known as "streaming video" which is considered a potential medium for future online advertising.

Nielsen/NetRatings reported that in December 1999, there were 4.7 million household broadband subscribers. By December 2000, that number had increased to 11.7 million.[7] The Strategis Group's January 2001 report, "Residential High-Speed Internet: Cable Modems, DSL, and Fixed Wireless," stated that 6.1 million households had broadband technology in 2000, up from less than 2 million in 1999. Strategis further projected 11.6 million households will be broadband households in 2001 and 36 million in 2005.[8] J.D. Power and Associates' report, "2001 Internet Service Provider Residential Customer Satisfaction Survey," reported 17.7 million broadband users in July of 2001 as compared with 8.0 million in July of 2000.

One drawback of broadband is that it is more expensive than traditional dial-up Internet service, with monthly subscription rates typically being $40 to $50. However, as more and more users move to higher speed, higher cost Internet service, they are likely to use the Internet for higher valued applications such as managing their finances. T.S. Kelly, director and principal analyst at Nielsen/NetRatings explains, "As early Internet adopters upgrade to broadband, the faster connections are changing their online habits. Faster speeds improve the

[7] "Study: Broadband Access Soars," CyberAtlas, February 8, 2001.
[8] Note that Strategis is citing household numbers whereas Nielsen was citing numbers of individuals.

overall online experience, encouraging broadband surfers to explore more sites and spend more time online. All this added activity benefits advertisers, E-commerce sites and content players."[9]

4. Business Internet Usage

By today, all large and medium sized businesses are connected to the Internet. Many of these have installed high-speed networks for faster and more reliable Internet access. There has also been considerable growth in Internet penetration among small businesses. According to a report by International Data Corp., "Internet Services to Small Businesses: Profiles of Portals, Aggreportals, Destinations,"[10] 5.0 million small businesses were online in 2000, up from 3.9 million in 1998. The number of small businesses with home pages increased from about 1.2 million to 2.1 million over the same period. The article also cites an Emarketer study, stating that in 2001, "78 percent, or 5.9 million, of all small businesses are connected to the Internet and nearly half have active purposeful Websites."

5. Online Advertising

In the past few years, the Internet has emerged as an advertising medium. Spending on Internet advertising would be expected to divert some direct mail advertising, though it would be expected that most of this impact would be felt in Standard (formally known as third-class) Mail. Still, about one-eight of First-Class letter mail is advertising, and is therefore subject to diversion to the Internet.

Table 8 presents annual Internet advertising expenditures, as reported by PriceWaterhouseCoopers (PWC), in association with the Internet Advertising Bureau (IAB). Table 8 shows the tremendous growth in Internet advertising over the past five years, rising from $266.4 million to $8,225.0 million. Annual growth rates have been very high, with the 2000 total nearly 80 percent greater than the 1999 total. At the same time, quarterly data show the much-discussed slowdown in Internet advertising. Spending actually fell from 2000Q2 to 2000Q3, and were essentially flat throughout all of 2000. The slowdown was due to many factors, among them the sharp decline in the stock market capitalization of many companies that had, in the past, made heavy use of Internet advertising, questions about the effectiveness of Internet advertising relative to other media, and a general slowdown in economic activity in general and advertising activity in particular which continues to the present day.

Table 8. Annual Internet Advertising Expenditures (in millions of dollars as reported by PWC/IAB)

	1996	1997	1998	1999	2000
Spending	$266.4	$906.5	$1,920.9	$4,604.4	$8,225.0
% Change		240.3%	111.9%	139.7%	78.6%

It should be noted that the PWC/IAB measures of Internet advertising have been subject to some criticism. Other analysts estimate somewhat lower values. For example, McCann-

[9] "Move to Broadband Changes How the Web is Surfed," CyberAtlas, August 22, 2001.
[10] Summarized in CyberAtlas' August 6, 2001 article, "Small Business Embraces Net, Shuns E-Commerce."

Erickson WorldGroup estimated that 1999 Internet advertising revenues were about $2 billion, or less than half the level presented by PWC/IAB.

C. Factors Limiting Technological Diversion

Several factors serve to slow the adoption of technological alternatives to the mail. The first is that for the most part, the Postal Service works. Mail is delivered, as it always has, at a cost that has not risen any more than prices in general. Each year, trillions of dollars of payments move through the mail system. Businesses have extensive investments in maintaining detailed computer databases with information sorted by mailing address. As such, there is a "if it ain't broke, don't fix it" view toward the mail and many Americans do not see a glaring need for electronic alternatives.

Second, many Internet activities have not yet been proven to be profitable. The economic landscape is littered with dot.com companies that have failed to deliver their promised new paradigm. Online advertising, for example, is not yet proven to be cost effective. Certainly, there is an awareness of the vast opportunity for marketing via the Internet, but the response rate for online advertising remains lower than for direct mail.

Related to the above is the fact that many of the projections of future technological activity are made by firms that may have a vested interest in hyping the very technologies they are analyzing. Consequently, using their information requires great caution. A firm's projection that, say, "streaming media" is the next big thing may be based as much in the desire to sell streaming media analysis services as it is a sober assessment of the future. Third, while computer and Internet penetration are rising, they are not yet universal as is the mail. One reason is cost, and Internet access among lower income households, though growing, remains below that of middle- and upper-income households. A second reason why Internet penetration is less than universal is that some people simply do not want it, regardless of cost. As a consequence, some people will never use electronic bill payment, never send or receive an E-greeting, or make an online purchase. And as long as there remains a sizable non-connected population, most businesses will have to continue to use the mail for at least some of their customers.

Fourth, even among those connected to the Internet, many will not use it to transmit personal or financial information. Concerns about the privacy and security of Internet transmissions are common. According to a 2001 survey by Statistical Research, Inc. (SRI), three-quarters of respondents were concerned about the information they provide online being shared in unauthorized ways, though 60 percent said that the existence of a Web site privacy statement/policy would reduce their concerns. The Pew Internet Project's survey of Americans with and without Internet access found that 87 percent of respondents were concerned about credit card theft online and 70 percent were worried about computer viruses that could wipe out personal computer files.

But in response to these concerns, firms are focusing on privacy and security issues, with security issues receiving the most attention. The Gartner group estimates that over the next decade there will be a ten-fold increase in the share of business revenues dedicated to information security. International Data Corp. (IDC) projects that worldwide revenue in the Internet security software will rise from less than $4 billion in 1999 to $11 billion in 2004. Whether these specific projections ever materialize in unclear, but it is important to remember that the industry has already proven that it is willing and able to commit significant resources to solving technology problems which would otherwise hinder acceptance of computer and Internet usage.

D. Y2K Impacts

A seemingly distant memory now, the Y2K computer problem captured the attention of businesses, consumers, governments and the media through much of 1999. The concern that there would be mass failures of computer systems due to difficulty with the change from year 1999 to year 2000 proved largely unfounded. People were not sent 100-year phone bills, electrical systems did not fail, banks did not lose all account records. As a consequence, many consumer fears were allayed regarding the reliability of computers. Had Y2K been associated with any of these calamities, consumer acceptance of online bill-paying and other computer-related financial activities would have been significantly delayed. Instead, the year 2000 appears to have provided a pent-up demand for these services by at least some consumers. Simply put, December 1999 was not the time to start online bill paying. To some extent, the growth in computer bill paying in 2000 shown in the next section of this testimony confirms that consumers are increasingly ready to use computers and the Internet for financial activities.

Another important, perhaps even more important, aspect of the passing of Y2K is that it has freed up tremendous amounts of resources for businesses to focus on post-Y2K activities. With that task accomplished, firms have resources that can be used to deal with security and privacy issues, as well as developments in online banking, advertising, and other Internet applications. Thus, the passing of Y2K set the stage for further expansion of technological alternatives to the mail.

E. An Example of Mail Diversion: Household Bill Payments

More than ten percent of First-Class letter mail is comprised of household bill payments. Traditionally, households have paid the vast majority of their bills by mail, with a much smaller proportion being paid in-person. In recent years, however, households have been using electronic alternatives to the mail to pay bills. These alternatives include bill payment by an automatic funds transfer, (AFT), by ATM, by phone, as well as by computer. Household Diary Study data provide evidence of the growing popularity of technological alternatives to bill payments by mail. Table 9 shows the percentage of households using various methods of bill payment from 1995 through 2001, based on Household Diary Study data.

Table 9 shows that over the past five years, technological methods gained in popularity, though some of the increase from 1999 to 2000 may be due to a slight change in the wording of the survey question. The table shows that most households use the mail for at least some of their bill payments, though this number has declined. More important is that according to the Diary Study results, more than forty percent of households use at least one kind of electronic method to pay their bills.

Table 9. Percentage of Households Using Different Methods of Paying Bills

Payment Method	1995	1996	1997	1998	1999	2000	2001
Mail	97.0%	96.5%	97.1%	96.3%	96.3%	95.4%	94.4%
In Person	39.0%	37.6%	37.7%	34.2%	34.3%	33.2%	31.4%
Phone	1.4%	1.4%	1.6%	1.6%	2.1%	7.5%	9.3%
Computer	0.2%	0.5%	0.7%	1.1%	1.7%	5.5%	8.8%
ATM	0.7%	0.8%	0.7%	0.6%	0.7%	3.1%	2.7%
AFT	17.2%	18.5%	19.8%	19.6%	19.3%	35.8%	36.5%
Any Electronic Method	18.6%	19.9%	21.3%	21.3%	21.9%	39.8%	41.2%

A second measure of the acceptance of technological alternatives for bill paying is the share of bills paid by each method. This provides a look at how intensively people are using each alternative, e.g., whether they make one bill payment automatically each month or several. Table 10 shows the share of bills paid by each method from 1995 to 2001 again using Household Diary Study data.

Table 10 shows that from 1998 to 2001, the share of household bills paid by mail has declined from 85.1 percent to 79.0 percent. This is equivalent to a loss of about 850 million mailed bill payments over a three-year period, just from households. Table 10 also shows that payments in-person have declined, suggesting that in the absence of growth in the use of electronic alternatives, the mail payment share might have increased over this time period. Instead, there has been a noticeable increase in the share of household bills paid using an electronic method, with almost one in seven bills household bills being paid electronically in 2001, according to the Diary Study data.

Table 10. Share of Household Bills Paid by Each Method

Payment Method	1995	1996	1997	1998	1999	2000	2001
Mail	85.5%	84.4%	84.8%	85.1%	84.4%	80.3%	79.0%
In Person	10.6%	10.5%	9.9%	9.5%	9.3%	7.5%	6.9%
Phone	0.4%	0.6%	0.5%	0.5%	0.5%	1.6%	1.8%
Computer	0.2%	0.4%	0.6%	0.7%	1.0%	2.8%	3.8%
ATM	0.1%	0.1%	0.2%	0.1%	0.2%	0.4%	0.3%
AFT	3.2%	4.0%	4.0%	4.3%	4.6%	7.5%	8.2%
Total	100.0%	100.0%	100.0%	100.0%	100.0%	100.0%	100.0%
Sub-Total Electronic Share	3.9%	5.2%	5.3%	5.5%	6.3%	12.2%	14.1%

Table 10 also shows an increase in the share of household bills paid by computer. The computer payment share is important because bill-paying by computer opens up a host of other possible avenues for mail diversion. Bills can be viewed (received) instead of mailed. Financial statements can also be viewed online instead of mailed. And should households begin receiving a large portion of their financial information across the Internet instead of the mail, it is likely that advertising mail would follow.

F. An Example of Mail Diversion: Internet Advertising

Internet advertising appears to have diverted direct mail advertising, though most of this impact is found in Standard, and not First-Class, Mail. Still, the impact of the Internet on direct mail shows how a new competitor can affect an existing business. Table 11 presents shares of total advertising expenditures for different media in 1990 and from 1995 through 2000.

A review of Table 11 shows that from 1995 to 2000 the Internet's share of advertising expenditures has increased from essentially zero to 3.5 percent of total advertising. Table 11 suggests that much of this increase in Internet advertising has come at the expense of direct mail. The decline in advertising share experienced by newspapers and non-cable television are part of long-term trends that pre-date the Internet. In 1980, for example, newspapers earned 27.6 percent of total advertising expenditures. In contrast, direct mail's share of total advertising expenditures in 1980 was only 14.2 percent, and it climbed steadily to 20.2 percent in 1995 before beginning its decline at the same time that Internet advertising began. Thus, Table 11 suggests that the emergence of Internet advertising has had an adverse effect on direct mail advertising.

Table 11. Advertising Expenditures Shares by Media

Media	1990	1995	1996	1997	1998	1999	2000
Direct Mail*	18.0	20.2	19.7	19.6	19.6	19.1	18.7
Newspapers	24.8	22.3	21.9	22.1	21.9	21.4	20.8
Magazines	5.2	5.3	5.1	5.2	5.2	5.1	5.0
Television**	22.4	20.1	20.6	19.6	17.7	17.2	16.6
Cable-TV	2.2	3.1	3.7	4.0	4.1	4.5	4.7
Radio	6.7	7.0	7.0	7.2	7.4	7.8	7.7
Internet	0.0	0.0	0.2	0.5	0.9	2.1	3.5
Other	20.7	22.0	21.8	21.8	23.2	22.8	23.0

(**sources:** McCann-Erickson, U.S. Postal Service, IAB, and author's calculations)
* Direct Mail advertising is predominantly sent as Standard Mail, not First-Class Mail
** Television, excluding cable-TV

IV. Measuring Technological Diversion

A. Conceptual Framework

Having described some key technological developments, it is time to create a framework for analyzing mail diversion. One way to understand technological diversion is to consider the activity of a single mailer. Suppose that in 1996, this mailer sent and received 300 pieces of First-Class letter mail. Further suppose that this mailer already makes use of some technological alternatives to the mail. A few faxes are sent or received instead of letters being mailed and a few bills are paid by phone or through an electronic funds transfer instead of being mailed. Altogether, suppose that these alternatives have reduced this mailer's volume of sent and received letters by 20. In other words, in the absence of these technological alternatives, the mailer would have sent and received 320 pieces of letter mail.

A first result, then, is that for this mailer, the "base volume" has been reduced from 320 to 300 pieces. Therefore, in the absence of any further technological uses, and for the sake of simplicity, in the absence of any changes in postal rates, economic activity or other drivers of mail volume, this mailer's 1997 volume of letter mail would also be 300 pieces. Thus, past diversion has already reduced 1997 volume from what it would have otherwise been.

It is possible that in 1997, the mailer would make less use of technological alternatives. Correspondences that were faxed in 1996 could be mailed in 1997, in which case technological diversion would be reduced and volume in 1997 would be greater than 300 pieces. But the reality is that technological use tends to grow over time. More correspondences are faxed as more and more people, businesses, and government agencies gain fax capabilities. More bills are paid electronically because more businesses accept electronic payment. In this typical case, the mailer would make greater use of technological alternatives in 1997 than in 1996. Ignoring other positive influences on volume, then this mailer may send and receive only 290 letters in 1997.

Thus, there are two measures of diversion in 1997 – the incremental diversion that occurred between 1996 and 1997 (10 pieces) and the cumulative diversion through 1997 (30 pieces consisting of the 20 that were diverted through 1996 and the ten additional diverted pieces in 1997).

Diversion is likely to grow over time for a given mailer, but this is not the only impact of technological alternatives on mail volume. A second impact is that over time, more and more mailers begin using technological alternatives. A second mailer, unaffected by diversion up through 1996 and sending and receiving 320 pieces of mail in that year begins using technological alternatives in 1997, reducing volume sent and received to 310. Therefore, incremental diversion for these two mailers in 1997 is 20 pieces and cumulative diversion through 1997 is 40 pieces.

Add to this process the rapid penetration of the Internet into households and businesses. E-mail becomes a substitute for correspondence mail. And as more and more households and businesses acquire Internet access, the impact is leveraged. Already "connected" individuals make greater use of E-mail as there are more other users with which to communicate. In addition, new users begin diverting letter mail as well. Finally, consider how computer/Internet access allows for greater range of technological uses, from downloading information that might otherwise be mailed, to E-filing of tax returns, to online banking, to the receipt of Internet advertising instead of direct mail. Businesses began to communicate over the Internet, reducing business-to-business correspondence, invoices, financial records, receipts, bills, and bill payments.

Eventually, the diversion process will be largely completed. At some future date, all the people who will be using technological alternatives will be using technological alternatives to their fullest and all the mail that will be diverted has been diverted. At this point, incremental (new) diversion is zero, but cumulative (total historical) diversion is quite large. This is likely to have been true for the telephone, which over a long period of time reduced letter volume, but at present has little additional negative impacts on the mail. In terms of the volume forecasting process, the diversion effects of the telephone need not be considered because the historical cumulative diversion is already present in (or more correctly, missing from) the base volume and there is little new incremental diversion to be considered.

In contrast, for technologies that are fairly new, in terms of the number of and past experience of current users, incremental diversion is large relative to cumulative diversion. While cumulative diversion is also present in (missing from) the base volume, future incremental diversion is not, and this diversion must be considered in the volume forecast.

Naturally, there are many other factors that affect mail volume and this is why volume demand equations rely on sophisticated econometric techniques designed to isolate the individual impacts of each factor. The key question is: where are we on the diversion process with respect to recent technological developments, most notably the Internet? Has most of the mail that will be diverted already been diverted, or are we at the beginning, or somewhere in the middle?

B. Possible Variables that Measure Technological Diversion

1. Number of Uses

One measure of technological diversion is the number of uses of various technological alternatives, e.g., the number of fax messages sent in a given year. This was the approach first used by the Postal Service in the early 1990s to estimate and project diversion of mail. Estimates of the number of faxes, E-mails, EFTs, and EDIs were combined with estimates of the diversion ratio -- e.g. the number of pieces of mail diverted for each use of a technological alternative -- to derive an estimate of diversion due to each technology.

Diversion ratios are a useful way of thinking about the problem. Importantly, diversion ratios are typically less than one, reflecting the understanding that not every electronic message replaces mail. But from an empirical standpoint, diversion ratio analysis has several drawbacks. First, reliable estimates of the number of electronic messages are difficult, if not impossible, to obtain. Various estimates of the number of E-mail messages sent in a year are found to differ by a factor of ten. Second, diversion ratios are difficult to estimate and earlier attempts to quantify these ratios were based on limited surveys and judgment. Given the large number of E-mail messages sent, small changes in the assumed diversion ratio could have large impacts on the estimated volume loss.

A third drawback of diversion ratio analysis is that it focused on individual technologies, i.e., a diversion ratio for E-mail, a diversion ratio for fax, a diversion ratio for electronic funds transfers, etc. The drawbacks of this approach are two-fold. First, from the standpoint of the mail, it hardly matters which technology has caused the diversion. It is of little consequence whether a piece of mail was diverted by a fax or an E-mail message. Second, many technologies replace other technologies, as opposed to replacing the mail. E-mail with attachment replaces fax messaging, for instance. As such, increases or decreases in the level of fax messaging may not have any direct impact on the mail.

2. Number of Users

A second possible measure of technological diversion is the number of users of certain technologies, as for example, the number of Internet users. Data on Internet users can be obtained with some reliability and could be included in the econometric equations to estimate the impact of increases in users on mail volume. But while the number of users measures the "width" of use, it does not measure the "depth" of use. In other words, a model including users as an explanatory variable would have the result that once the number of users stops growing (reaches its saturation level), diversion stops as well. This is not likely because individual users will tend to make greater use of technological alternatives over time. To claim otherwise is to suggest that once a individual becomes a user of a technology, all of the mail diversion occurs in the year of initial use, with none occurring in future years, which is clearly not the case. Accordingly, number of users is not the right variable to measure technological diversion.

3. Dollars of Use

The Postal Service currently uses measures of dollars spent on technological use as the driver of the electronic diversion of mail. Expenditures reflect both changes in "width" (number of users) and "depth" (intensity of use). Accordingly, two Internet expenditure measures are included in the econometric analysis of mail volumes. Household expenditures on Internet Service Providers are included in the demand equation for single-piece letters. Internet advertising expenditures are included in the demand equations for Standard Mail.

There are some possible drawbacks to using expenditures variables. Expenditures are a price times quantity variable and changes in these two components of expenditures could have different effects on mail volume. For example, Internet advertising expenditures could decline because less Internet advertising is bought or because the price of Internet advertising has declined. However, following the principle that "you get what you pay for," the expenditure measure still serves its purpose. A decline in the price of Internet advertising is likely to result from a decline in the medium's effectiveness, which in turn, would make it less attractive relative to direct mail. For the same reasons, increases in expenditures due to price increases would be expected to increase technological diversion on the grounds that people are paying more to get more. And expenditure measures have one key advantage; they allow for aggregation across different products using a common measure – the dollar.

C. Household Expenditures on Internet Service Providers

1. Overview

Household expenditures on Internet Service Providers (ISP) is a broad measure of technological activity. The US Department of Commerce reports monthly levels (expressed an annual rates) for household ISP expenditures, beginning in January 1988. As of April of 2001, the last monthly value used in the volume forecasts presented in the R2001-1 rate case, household ISP expenditures had climbed to an annual rate of $25 billion, or about $2 billion per month.

Internet Service Providers are companies that provide Internet access, most notably America Online (AOL), Microsoft Network (MSN), Earthlink and Compuserve. According to the Online Census from Telecommunications Reports International (TRI), the total number of US subscribers to Internet Services rose to 70.7 million in the second quarter of 2001, a figure which is also reported as the number of households subscribing on the grounds that there is usually only one "subscriber" per household. In other words, several household members may be sharing a single subscription, although the distinction between households and individuals is not always made clear.

Increases in household ISP expenditures capture many of the technological developments discussed in the previous chapter. First, growth in expenditures reflects growth in the online population. Second, it reflects increases in dollars spent per person online (also referred to as a "user"). Growth in dollars per user reflects movements toward higher cost, higher speed connections (such as broadband) as well as increased value-added features that provide, among other things, increased reliability and security. Increases in dollar per user also reflect a greater intensity of use since many ISPs have different rate structures based on monthly usage, with the highest monthly fees being charged for unlimited monthly access. Thus, increases in ISP spending are reflective of increases in the number of users, increases in the quality of their service, and increases in their time spent online.

The ISP expenditure variable has some drawbacks as an explanatory factor for mail volumes. First, there is the basic issue of the degree to which Internet activity diverts mail volume, an issue that will be addressed presently. Beyond this fundamental question, there are issues related to the ISP variable as a measure of Internet activity. First, not all Internet users are paid subscribers to online services. Some people have free Internet service, though the number of free services is shrinking rapidly. According to TRI, the number of free Internet subscribers declined 11 percent from quarter 1 to quarter 2 of 2001, and now number barely one in ten total subscribers. For the most part, then, ISP expenditures capture Internet activity by households.

The ISP variable does not measure Internet service expenditures by non-households. Nonetheless, it seems likely that growth in household ISP expenditures correlates closely with increases in business Internet expenditures, perhaps with a lag reflecting the somewhat slower adoption of the Internet by households as opposed to businesses. Still, it is probably the case that increases in household ISP expenditures correlate with increases in Internet activity in the non-household sector, and therefore the variable serves as a proxy measure of non-household Internet activity.

Taken a step further, ISP expenditures may serve as a measure of overall technological acceptance in the US. The growth in Internet use is not an isolated development. It is interwoven with a general growth in the use of technology by individuals, and the expansion of networking between individuals, businesses, and the government.

The foregoing issues aside, the main issue is whether household ISP expenditures acts as a driver of mail diversion. Conceptually, this seems reasonable. Access to the Internet provides users with new alternatives to the mail. Correspondence mail can be replaced by E-mail, and while the level of household-to-household correspondence mail is not large, E-mail is also used as an alternative to correspondence between households and nonhouseholds. Internet access also allows for greater opportunities to divert financial mail, including bill payments and presentments, along with financial statements. As people have the ability to check their financial accounts online, there is less of a need for mailed information. Third, increases in ISP expenditures may reflect a general trend toward greater technological activity, activity that is not necessarily directly related to the Internet, but is consistent with the greater household acceptance of technology, and with that, greater use of technological alternatives to the mail.

2. Looking at the Data

Chart B presents household ISP expenditures from January of 1988 to April of 2001. Growth has been rapid. It was not until 1994, that ISP expenditures reached an annual rate of \$1 billion. Expenditures quadrupled by 1998, reaching \$4 billion, and have since increased by a factor of five, now running at a rate of more than \$20 billion per year. Closer examination of Chart B shows that in mid-1999, ISP expenditures began a rapid period of growth, somewhat distinct from the prior period.

Chart B
Household Expenditures on Internet Service Providers
($ billions)

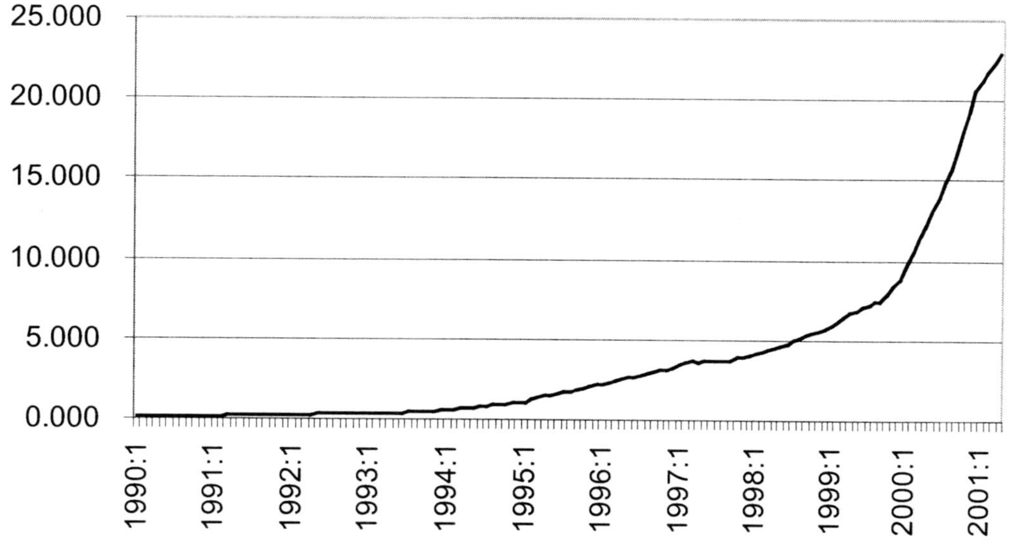

3. Modeling ISP Expenditures

Projected ISP expenditures are modeled as being equal to the product of two key drivers of ISP spending – increases in the number of active Internet users and increases in the number of dollars spent per user. The ISP expenditure model does not include macro-economic variables such as real changes in GDP or consumption. This is not to say that overall economic conditions have no influence on ISP spending, indeed they do and the impact may be large. It is merely an acceptance of the fact that a variable that has increased by a factor of ten over the past five years, and quadrupled over the past two years, is not well-modeled by economic variables that typically grow less than five percent per year.[11]

ISP expenditures are modeled using monthly data beginning in April 1999. This is done for two reasons. First, it is around this time that there occurred a break in the data, and ISP expenditures began to increase substantially. Second, in April 1999, Nielsen/ NetRatings began reporting a monthly series on "active Internet users" defined as people who were online in the prior month. As discussed earlier, this is a better measure than the total population with Internet access, since many people with access rarely connect to the Internet. It is also important to recognize that while use within the past month might seem a low standard for being an "active" user, the typical user is in fact quite active, with average hours of usage ranging from seven to ten hours per month over the sample period.

[11] Recent work includes total consumption expenditures as part of the model used to project ISP expenditures, which have been substantially revised by the Commerce Department.

Conceptually, one could use "ISP subscribers" as opposed to "active users" in this analysis. However, subscriber data is not presented on a monthly basis and there is, as mentioned earlier, some uncertainty regarding whether "subscribers" refers to individuals or households. The Nielsen numbers are monthly measures of individuals who are active users of the Internet at home.

Chart A showed active users from April 1999 to May 2001. Active users increased from 57.6 million to 103.0 million. Growth was fairly steady, averaging nearly two million new users each month.

4. Statistical Model of Active Users

a. Bass Curve Model

Active users are modeled by estimating a Bass curve also known as a market penetration curve or product diffusion curve. A Bass curve is a line showing how much of a potential market has adopted a new product as a function of time and the number of previous adopters.[12] The formula is given below:

$$N_T = N_{t-1} + p(-N_{t-1}) + q\left(\frac{N_{t-1}}{m}\right)(m - N_{t-1})$$

where N is the number of adopters at time "t" or in the previous period (depending on the time subscript), ***p*** is a parameter for the force of outside or exogenous factors affecting adoption (i.e. factors independent of the number who have adopted the innovation previously, sometimes referred to as the coefficient of innovation), **q** is the parameter for the force of previous adopters in creating new adopters (e.g. by word of mouth, sometimes called to coefficient of imitation), and ***m*** is the maximum size of the market or ceiling value The Bass equation models adoption as a function only of time given these parameter values.

The logic of the Bass curve is that there are two channels for the diffusion of information that prompts people to adopt a new product. One is the exogenous influence, such as advertising or an increase in demand through greater numbers of complementary products. Regardless of the number of people currently using the product, a certain number will adopt in each time period due to these influences. The second influence is the endogenous effect, which is the effect of the previous adopters. By word of mouth, or by greater usefulness if the products interact positively (for instance telephones, which are more useful the more other people own one), demand for the new product increases as the number of users itself increases. Product diffusion never ceases, but diminishes to insignificance as the ceiling ***m*** is approached. This parameter allows for the existence of permanent non-adopters, that is the ceiling can be less than total population. It should be noted that no other factors influence the adoption rate in this simplified model; product price, for instance, is not included. Therefore, among the total population of potential adopters (defined by ***m***), the percentage of people who adopt in a given period, given that they have not yet adopted, is proportional to the percent of people who have already adopted.

[12] Mahajan, V., Miller, E., Bass, F.M., "New Product Diffusion Models in Marketing: A Review and Directions for Research," *Journal of Marketing* **54**, 1 (January 1990), pp. 1-26.

Given a set of parameter estimates, the number of adopters can be forecast for the next time period, and this number can be used to solve for the number in the following period, giving an extrapolation as far into the future as is desired. The pattern generated by the Bass curve is an "S curve," with an initial period of exponential growth, followed by continuously diminishing growth rates with an asymptotic approach to the ceiling value.

A Bass Curve is estimated for the historical data on active users, with monthly values beginning in April 1999 through May 2001, a total of 26 observations. The resulting Bass curve parameters are presented below in Table 12.

Table 12. Estimated Bass Curve Parameters for Active Internet Users

Bass Curve Parameter	Estimated Value
m (ceiling on active users)	306.70 million
p (coefficient of innovation):	0.006322
q (coefficient of imitation):	0.002059

Reviewing the results presented above, the ceiling on active users is estimated to be 306.70 million people. At first, this result seems unreasonable – US population today is about 280 million people. But the ceiling is a future value, achieved years into the future when US population will be much greater. More generally, then, the high ceiling value merely suggests that at some time Internet access may become universal within the US, much like the telephone or the television. (In fact, eventually the Internet may be your television).

More relevant is the value of active users in the next few years. Table 13 shows active users in May of 1999, 2000, and 2001, with projections in May of 2002, 2003, 2004, and 2005. Annual growth rates are also presented, which show the gradual slowing of growth that is typical of market penetration curves.

Table 13. Actual and Projected Values of Active Internet Users

Month	Active Users	Annual Growth
May 1999	61.40 million	
May 2000	82.70 million	34.7%
May 2001	103.01 million	24.6%
May 2002	122.06 million	18.5%
May 2003	139.27 million	14.1%
May 2004	154.84 million	11.2%
May 2005	168.90 million	9.7%

5. Dollars per Active User

The second part of modeling household ISP expenditures is to model dollars spent per active user, which when multiplied by projections of active users will yield a projection of total ISP spending. Dollars (of ISP expenditures) per active user are calculated by dividing ISP expenditures by 12 (to convert the annual rate to a monthly value) and then dividing the result by the number of active users. Historical data on dollars per user are presented in Chart C from April 1999 through April 2001, a total of 25 observations.

Chart C shows that dollars per active user approximately doubled over a two year period, rising form \$9.67 per month in April of 1999 to \$18.61 per month in April of 2001. In fact, from April through December 1999, dollars per user was roughly constant, with the strong growth beginning in January 2000. This is consistent with the view presented earlier that the passing of the Y2K computer problem without major incident may have contributed to the public's desire for greater technological activity.

Chart C
Dollars per Month of ISP Expenditures per Active Internet User

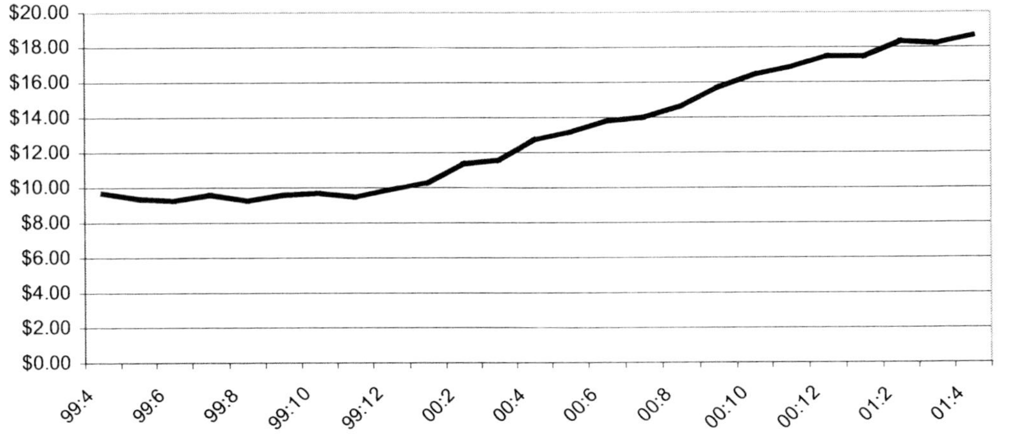

Dollars per user are modeled as a simple time trend, using all 25 monthly observations as follows:

$$\$/\text{User} = \alpha + \beta t + \mu$$

where α is the intercept (value at time zero), β is the estimate time trend value, t is month running from 0 to 24, and μ is the regression residual. The regression results, including estimated coefficients, standard errors, and t-statistics are presented in Table 14.

Table 14. Results of Regression of Dollars per Active User per Month

Observations	**25**		
Adjusted R-squared	0.9830		
F-Statistic	364.2		
Coefficient Estimates	Value	Standard Error	T-Stat
Intercept	\$7.5985	0.3339	22.76
Time Trend	\$0.4551	0.0239	19.08

The regression results show that the average monthly increase in ISP spending per active user is about 45.5 cents. A constant monthly increase will have the effect of resulting in a declining percentage increase. Table 15 shows actual and projected monthly values from April of 1999 through April of 2005. Annual percentage changes are also presented, which show the eventually declining growth rate that is expected to occur.

Table 15 shows that by April of 2005, active Internet users are projected to spend about $40 per month on Internet services. This is somewhat more than twice the average monthly expenditure in April of 2001 and is consistent with the view that over time more and more users will be switching to higher speed, higher cost Internet services that include many value-added features such as enhanced security and greater reliability.

Table 15. Actual and Projected Monthly ISP Spending per Active Internet Users

Month	Active Users	Annual Growth
April 1999	$9.67	
April 2000	$12.69	31.2%
April 2001	$18.61	46.7%
April 2002	$24.07	29.3%
April 2003	$29.53	22.7%
April 2004	$35.00	18.5%
April 2005	$40.46	15.6%

6. Comparison of Actual with Model of ISP Expenditures

Multiplying ISP expenditures per active user per month by monthly active users yields estimates of total monthly ISP expenditures.[13] These estimates (or fitted values) are compared to actual values from April 1999 through April 2001 in Chart D. Analysis shows that the fitted values closely match the actual values, particularly in the last year of data. The modeling of past ISP expenditures is therefore used to make projections of future spending.

Chart D
Comparison of Actual and Fitted ISP

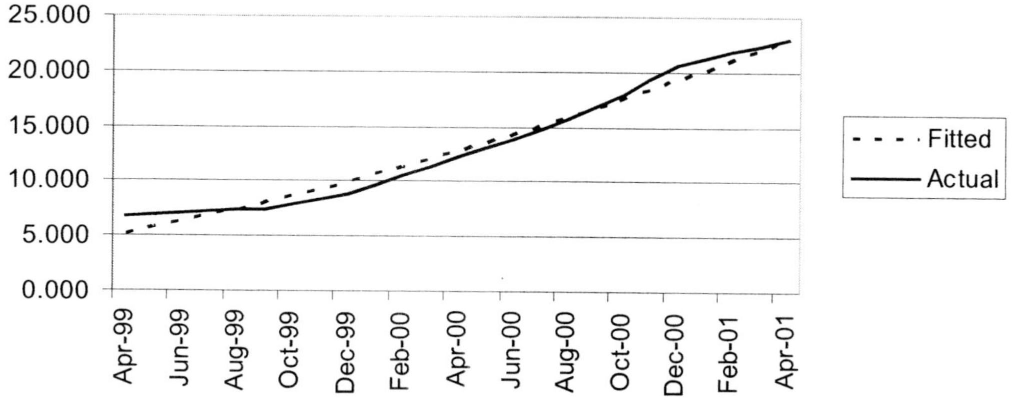

[13] Monthly expenditures are multiplied by 12 to create expenditures at an annual rate, consistent with how the ISP expenditure variable is reported.

7. Projections of Future ISP Expenditures

Future ISP expenditures are projected by multiplying projections of expenditures per active user per month by projections of the number of active users. These monthly projections served as inputs into the volume forecasts presented by George Tolley in the R2001-1 case. Historical and projected values for the month of April in 1999 through 2005 are presented below in Table 16.

Table 16. Actual and Projected Values of Household ISP Expenditures

Month	ISP Expenditures	Annual Growth
April 1999	$6.686 billion	
April 2000	$12.224 billion	82.8%
April 2001	$22.885 billion	87.2%
April 2002	$34.822 billion	52.2%
April 2003	$48.875 billion	40.4%
April 2004	$64.503 billion	32.0%
April 2005	$81.456 billion	26.3%

The projection is that household ISP expenditures will rise from $22.885 billion per year in April of 2001 to $81.456 billion per year in April of 2005. It should be recognized that this future number is both a quantity and quality value. In other words, it is projected that households will purchase the equivalent of $81.456 billion worth of ISP services in 2005. To put this in perspective, one might have looked at increases in household spending on computers and concluded that households would be spending twice as much on computers in 2001 than in 1998. In fact, the price of a home computer has actually declined since that time. But, in terms of purchasing power, spending has doubled in that the $1,500 computer of today has twice the power of a $1,500 computer purchased in 1998. A similar result might occur with ISP spending.

V. Measures of the Impact of Technology on Letter Volume

A. Negative and Positive Impacts

First-Class letter volume is often decomposed into two separate categories: single-piece letters and bulk (or workshare) letters. Workshare letters are large volume letter mailings in which the mailer presorts and automates the mail and receives a postage discount in return for the cost savings that presortation and automation provides to the Postal Service. Chart E shows fiscal year volumes of both single-piece and workshare letters from 1996 through 2001.

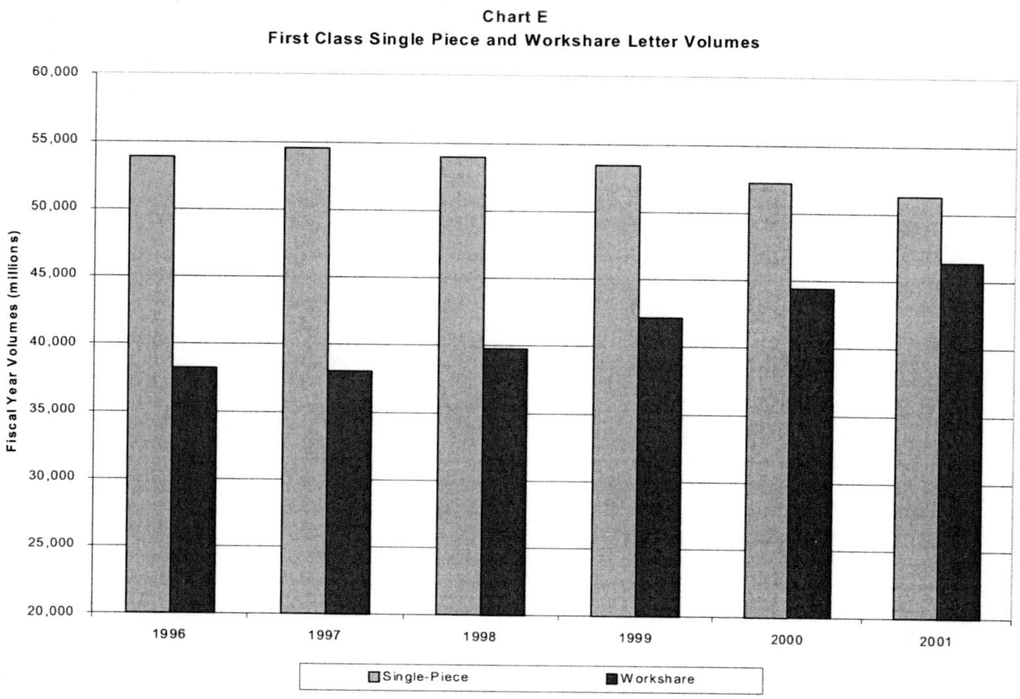

Earlier in this chapter, it was noted that technological changes have had both negative and positive impacts on mail volume. Workshare letter volume has probably benefited from technological advancements that have lowered mailer presort and automation costs. In addition, mass marketing databases have made direct mail advertising more effective, which has positively affected the use of First-Class workshare letters as advertising mail. Chart E shows that volume has increased over the past five years from about 38 billion pieces to 46 billion pieces.

With respect to single-piece letter volume, however, technological change has had a clear negative impact. Volume declined from about 54 billion pieces in 1996 to 51 billion pieces in 2001. This volume decline was somewhat unexpected, given a period of relatively strong economic growth and moderate increases in letter postage rates, and the Postal Service econometric models that existed at that time consistently overestimated single-piece letter volume.

In the R2001-1 rate case, the volume equation for single-piece letters included household expenditures on Internet Service Providers as an explanatory variable. The econometric work showed that this variable has a statistically significant negative impact on the volume single-piece letters. The estimated impact of this variable on single-piece volume was presented in Dr. Tolley's R2001-1 volume forecast testimony, reprinted below as Table 17. The econometric estimates were performed by Dr. Tolley's colleague at RCF, Thomas Thress who also testified to these equations in the rate case.

Table 17 shows that there are many other factors that affect single-piece letter volume, including real own-price, the price of First-Class cards (which can act as a substitute for letters), the workshare discount (the discount granted to senders of bulk First-Class letters), as well as income and population measures.[14] The equation also shows the impact of a postal rule change in 1995 which temporarily encouraged the use of single-piece mail. Finally, the model includes a negative trend term that explains an 8.86 percent decline in single-piece letter volume over the past five years. The time trend effect is a continuation of longer term negative influences on single-piece letter volume including the gradual decline in household letter writing and the use of long-standing alternatives to the mail that predate the widespread adoption of the Internet. Finally, Table 17 also shows that increases in ISP expenditures were responsible for an 8.57 percent decline in single-piece letter volume

Table 17. Contributions to Change in Single-Piece First-Class Letter Volume for the Five-Year Period Ending 2001Q3

Variable	Estimated Elasticity	Estimated Volume Impact in past 5 years
Real Own-Price	-0.311	+2.32%
Real Workshare Discount	-0.027	-0.54%
Real Price of Post Cards	0.004	-0.04%
Real Income per Adult	0.512	+5.10%
UCAP	0.099	-0.26%
Adult Population	1.000	4.50%
1995 Rule Change		3.58%
Long-term Trend		-8.86%
ISP Expenditures	-0.498	-8.57%
Total		-3.80%

Source: Testimony of George Tolley, United States Postal Service, R2001-1

In contrast to single-piece volume, workshare volume has continued to grow, increasing by more than 20 percent over the same five-year period considered in Table 17. To some extent, the differing volume experiences are two sides of the same coin -- mailers have shifted volume from single-piece to workshare in response to increases in the discounts granted to workshare mailers. But beyond this shifting of mail, the evidence is that workshare mail volume has benefited from technological advancements that have reduced mailers' costs of presorting and automating their mail. This is not to say that workshare letter volume has not been subject to some negative impacts due to the growing use of the Internet as a communication alternative. Instead, the view is that, with respect to workshare volume, the negative impacts of technological change have up to this point been far less than the positive impacts.

B. Summary of Recent Impacts of Technology on Letter Volume

Thus, technological changes have had mixed effects on letter volume. The Internet and related technologies have reduced volume while other technological advancements have lowered costs for bulk mailers. The magnitudes of these impacts, both historically and

[14] The econometric analysis uses volume per adult as the dependent variable, giving adult population an implicit volume elasticity of 1.0. Moreover, the ISP expenditure variable was entered in the equation in absolute terms,

forecasted, are shown in Table 18. Historical estimates are based on the results of econometric analysis. Forecast impacts are based on the projections of future ISP expenditures (presented earlier in this chapter), as well as the assumption that mailing costs will continue to decline as they have in the past.

Table 18 shows that in the last 1980s and early 1990s, the considered technological changes had a net positive impact on letter volume. Diversion of mail due to Internet use was relatively small, with about three-quarters of a billion pieces, or less than one percent of total letter volume, having been lost through 1992. Over that same period, reductions in mailing costs are estimated to have been responsible for an even greater increase in letter volume (+1,619 million pieces) so that the combined effects of technological change through 1992 was a net increase of 850 million pieces.

After 1992, however, the negative impacts of the Internet began to outweigh the positive impacts of declining mailing costs. Through 1996, the net effect was still slightly positive, but has since turned decidedly negative. From 1996 through 2001, it is estimated that about 7.5 billion pieces of letter mail were lost due to growth in the use of the Internet and related technologies. During this period, declining mailing costs are estimated to have added about 3.6 billion pieces, but the net effect of technological change through 2001 is estimated to be a reduction of about 3.9 billion pieces of letter mail.

Table 18. Impact of Technological Change on First-Class Letters Volume (millions of pieces, cumulative since 1987)

Year	*Internet*	*Mailing Costs*	*Total*
1988	(102)	318	216
1989	(283)	647	364
1990	(398)	949	551
1991	(603)	1,289	686
1992	(769)	1,619	850
1993	(892)	1,606*	714
1994	(1,211)	1,925	714
1995	(1,763)	2,273	510
1996	(2,405)	2,505	100
1997	(2,896)	2,430*	(466)
1998	(3,253)	2,731	(522)
1999	(3,939)	3,032	(907)
2000	(5,241)	3,320	(1,921)
2001	(7,480)	3,602	(3,878)
2002	(9,163)	3,872	(5,291)
2003	(10,808)	4,140	(6,668)
2004	(12,305)	4,407	(7,898)

* Postal Service rule changes caused temporary increases in mailing costs

Sources: Testimony of Thomas Thress, United States Postal Service, R2001-1
Testimony of George Tolley, United States Postal Service, R2001-1
Testimony of Peter Bernstein, United States Postal Service, R2001-1

as opposed to in log terms like the other variables.

Moving forward, the negative impacts are forecasted to grow considerably. From 2001 through 2004, the loss of letter volume due to the Internet is projected to increase from 7.5 billion to 12.3 billion pieces, a net loss of 4.8 billion pieces over this period. Declining mailing costs will add only about 800 million pieces over this period, with the cumulative impact growing from 3.6 billion to 4.4 billion from 2001 to 2004. As such, the forecast is that technological changes will reduce First-Class letter volume by about four billion pieces from 2001 through 2004. In terms of revenue, this is equivalent to a loss of about $1.5 billion.

C. Longer-Term Volume Projections

Although not required for a postal rate case, longer-term volume projections are often made to aid Postal Service management with planning decisions. These projections rely more heavily on judgment than econometric analysis, since many of the drivers of volume a decade in the future cannot be well predicted at this time. This is especially true for measures of mail diversion. Table 17 below presents projections of letter volume in 2010. It is based on projections of volume of different components of First-Class letters, in the absence of new diversion, along with projections of new diversion. The no-diversion growth rates are based on analysis of historical trends, with financial mail (bills, bill payments, and statements) growing with real GDP, advertising mail growing somewhat more rapidly, and other mail growing less rapidly, with household correspondence mail increasing with population. The diversion forecast is that eight percent of non-financial letter mail will be lost to electronic alternatives over the next eight years. Financial letter mail (bills, bill payments, and statements) is expected to be more affected, due to growing use of bill payment alternatives and computer presentment of bills and statements.

Table 19. A Forecast of Letter Volume in 2010

Type of Letter Mail	2002 Estimated Volume	No Diversion Growth Rate	Projected No Diversion 2010 Volume	Projected New Diversion to 2010	Projected 2010 Volume
HH to HH					
Correspondence	6,784	1.0%	7,346	8%	6,758
NHH to HH					
Bills and Statements	21,805	2.7%	26,985	25%	20,239
Advertising	7,268	3.6%	9,645	8%	8,873
Other	14,537	2.0%	17,032	8%	15,669
HH to NHH					
Bill Payments	10,660	2.7%	13,193	25%	9,895
Other	1,938	2.0%	2,271	8%	2,089
NHH to NHH					
Bills and Statements	19,382	2.7%	23,987	25%	17,990
Advertising	4,845	3.6%	6,432	8%	5,917
Other	9,691	2.0%	11,355	8%	10,447
Total Letter Mail	**96,911**	**2.5%**	**118,243**	**17%**	**97,877**

Key Economic Assumptions: No change in real postal rates. 2.7 percent annual average growth in real GDP

The long-term forecast shown in Table 19 is that in the absence of new diversion, letter mail volume would grow at an annual rate of 2.5 percent, reaching 118.2 billion pieces in 2010. Projected losses due to technological alternatives, affecting financial mail most heavily, are expected to reduce volume by 17 percent, to 97.9 billion pieces. This is only about one billion pieces greater than actual 2002 volume.

Clearly, there is considerable uncertainly about a forecast eight years into the future, Still, at present, it appears likely that continued growth in the use of alternatives to the mail will cause letter volume to remain close to its current level for the remainder of the decade.

D. Postal Service Response to Growth in Technological Alternatives

The development of the Internet and the greater use of technological alternatives to the mail have changed the competitive position of the Postal Service. No longer can the agency rely on increases in First-Class letter mail because of the absence of suitable alternatives. But increased competition can have advantages for the Postal Service, as it creates the impetus for change by a company that has long pursued a "business as usual" operating mode.

Among the actions that the Postal Service can take in response to technological competition are the following:

1. Gain Better Understanding of Historical and Likely Future Diversion

A first response of the Postal Service to the growing threat from technological alternatives to the mail is to continue the work described in this chapter and gain a better understanding of the empirical impacts of diversion. Declines in letter volume are easily measured; the reasons for those declines are not. This work is especially important in the current environment as a number of different factors – recession, recent rate increase, terrorism, and technological competition – have combined to reduce the use of the mail. Recessions eventually end, rate increases can be managed by the Postal Service, terrorism (hopefully) will be a one-time event. Volume losses to these events will likely return, and accurate estimates of the volume elasticities to economic factors will provide measures of that recovery and provide a useful baseline for analyzing the remaining impacts of technological spread.

Reliable long-term volume forecasts are crucial to the Postal Service because, by habit and by construction, it is slow to respond to change. Unlike many private sector companies, the Postal Service cannot easily decrease its scale of operations if revenues decline. The evidence suggests that the impacts discussed in this chapter will be persistent but gradual, allowing the Postal Service an opportunity to change.

2. Keep Rate Increases at or Below Inflation

Although the demand for First-Class letter mail is inelastic, it is not completely unaffected by changes in postage rates. Single-piece letter mail has an estimated price elasticity somewhat greater than 0.3 in absolute terms. Workshare letter mail is less elastic but overall, volume can be expected to decline by about one-half percent for each one cent increase in the letter rate. Given that letter volume has grown at an average rate of only one percent per year over the past decade, a one-half percent volume impact of price is important.

Still, the Postal Service must be careful not to control increases in the basic letter rate by allowing greater increases in the rates of other, more price sensitive mail categories.

3. Use Technology to Increase Contribution

Technological developments need not have only negative impacts on the mail. One plan being developed at the Postal Service is to use technology to develop "Intelligent Mail" that can provide value-added features to mailers and increase revenue. For example, technology can allow mailers to use mail bar codes to track their mail through the system. This could be especially valuable to billers who could better manage their flow of funds received and to advertisers who could better measure the timing between mail receipt and customer action. In this case, advertisers might be inclined to shift from Standard Mail to First-Class Mail, leading to net increases in contribution to overhead.

4. Control Costs

Finally, the Postal Service must continue the progress it has made in controlling costs. Again, technology can have a positive effect, as increased mail automation reduces labor cost. Still, about three-fourths of postal costs are labor-related, and the mostly unionized postal labor force is reluctant to grant wage or employment concessions. Ironically, recognition of the spread of technological alternatives to the mail has helped the Postal Service management in its negotiation with the various postal unions. In 2002, labor arbitration judges granted only moderate wage increases as part of the current three-year contract, citing increased technological competition and decreasing revenue assurance as key factors in that decision.

VI. Epilogue

On the morning of September 11, the Postal Service Board of Governors voted to officially file for a rate increase two weeks hence. Moments after that vote, the remainder of the meeting was canceled after news that a plane had crashed into the Pentagon, just a few miles away. The attack of September 11, and the worsening economy combined with the sending of anthrax through the mail caused mail volume in fiscal 2002 to decline by more than five billion pieces. First-Class letter mail declined more than one billion pieces, and the Postal Service found itself with unexpectedly large financial losses due to declining revenues and rising security costs.

It is difficult to determine the extent to which the 2002 volume losses were due to terrorism, economic decline, or the growing use of technological alternatives to the mail. In some ways, the events may have slowed the movement toward these alternatives. The Internet advertising market, for example, collapse with 2002 ad revenues expected to be at least twenty percent below the 2000 peak. Technology spending also fell sharply, possibly slowing the adoption of Internet by businesses and households alike.

Another complicating event occurred in July 2002 when the Commerce Department substantially revised its historical estimates of household spending on ISP services. This revision appears to have been related to the Department's estimate of "quality improvements" which had previously been factored in as part of an increase in spending. Table 20 presents a comparison of the ISP spending figures shown in Tables 15 and 16 with the values resulting from the revised Commerce Department estimates of ISP spending.

Table 20. Comparison of Original and Revised Estimates of ISP Spending

Month	Original ISP Estimates		Revised ISP Estimates	
	Expenditures (per year)	$ per User (per month)	Expenditures (per year)	$ per User (per month)
April 1999	$6.686 bln	$9.67	$6.102 bln	$8.83
April 2000	$12.224 bln	$12.69	$7.984 bln	$8.29
April 2001	$22.885 bln	$18.61	$10.672 bln	$8.68
April 2002	*$34.822 bln*	*$24.07*	$12.432 bln	$8.59

Italics indicate forecasted value based on original ISP spending estimates

ISP spending, originally estimated to have increased by 240 percent from April 1999 to April 2001 is now estimated to have rising just 75 percent. Put differently, the revised actual spending for April 2001 is less than half the originally estimated spending level. Furthermore, the revision of total ISP spending meant that spending per Internet user per month is now estimated to be essentially constant over the time period shown, as opposed to consistently increasing as was the case with the original data.

Other indicators of technological use are more consistent with the analysis presented in this chapter. The Postal Service's 2002 Household Diary Study found that the share of bills paid by mail fell to 75.3 percent in 2002, down from 79.0 percent in 2001, implying a loss of more than 400 million pieces of mail. The study also found that one in eight households regularly pay at least some bills online and, among these households, pay fewer than half their bills by mail.

Internet advertising appears to be evolving away from the banner ads of 2000 and toward mass e-mail advertising messages, almost costless to send. Workshare letter volume appears to be finally experiencing a net negative impact of technological change, with 2002 volume growing barely more than one percent.

Against this negative backdrop, the Postal Service received unexpectedly good fiscal news from a GAO investigation that discovered the Postal Service pension fund was in far better financial condition than previously thought. Proposed changes in future funding could reduce postal expenses by more than $2 billion per year, possibly allowing the Postal Service to delay any future rate increase until 2006.

Whether temporary rate stability will turn the tide toward electronic alternatives to the mail is uncertain. First-Class letter mail is relatively price inelastic, meaning that stable postal rates will do little to raise volume. The price of postage may have only a small impact on the decision to pay and receive bills by computer, for example. As a result, longer-term projections of First-Class letter volume remain largely the same as presented earlier in this chapter.

The Postal Service has faced change before. Over the past generation, it has been transformed from a heavily subsidized government operation to a break-even business more than doubled in size. The mail has transformed as well, from a medium commonly used for household communication to one in which more than 95 percent of the mail is sent by or received by non-households. And the threat of technological change is not unique to the current period. Over the past 200 years, new and then newer communication technologies have replaced the mail – telegraph, telephone, fax and e-mail. But these new technologies also replaced their predecessors, while mail volume increased. Today, telegraphs are never sent, phones are no longer connected to houses, and e-mail may soon make the fax machine obsolete. The mail will keep coming.

Chapter 6

BROADBAND ADOPTION IN THE UNITED STATES: AN EMPIRICAL ANALYSIS*

David E. Burnstein
LECG, Inc.
Debra J. Aron
Northwestern University
LECG, Inc.

I. INTRODUCTION

There is growing concern over the rate at which the Internet and broadband access to the Internet are being adopted in the U.S. By historical standards of new product adoption, Internet and broadband adoption have been nothing short of extraordinary. Nevertheless, these rates have fallen short of previous expectations and have even declined in the last few years.[1, 2] As a result, a policy debate has erupted over how to facilitate consumer adoption of the Internet and, particularly, broadband technology to access the Internet.[3] At its base, the

---

* The authors gratefully acknowledge the helpful comments of Larry Manheim, Eugene Burnstein, and the Editor on earlier drafts. All errors or omissions are our own, and the views expressed herein are those of the authors and not necessarily those of LECG or its individual directors or employees.

1 Based on an annual survey by Harris Interactive, the percentage of adults with access to the Internet (be it from home or work) increased from 9 percent in 1995 to 63 percent by the end of 2000, whereas since then it has only increased 4 percentage points to 67 percent by the end of 2002. "The Harris Poll: Online Population Continues to Expand but Just Barely," *Wall Street Journal Online*, February 5, 2003.

2 In June and December of 2000, the (trailing six month) percentage growth in broadband subscribership was approximately 60 percent; in subsequent reporting periods the (trailing six month) percentage growth had declined from 36 percent in June 2001 to 27 percent in June 2002. FCC (2002), Table 1.

3 See, for example, Jim Landers, "Broadband believer: FCC's chair presses for big leap for U.S. high-speed Internet," *The Seattle Times*, October 21, 2002 ("One of the few points of agreement in the broadband debate is that the move to high-speed access will remain slow till the FCC clears away a dense fog of regulatory ambiguity."); and Charles H. Ferguson, "The U.S. Broadband Problem," The Brookings Institution, Policy Brief #105, July 2002 ("The pace of deployment and technological progress in broadband, or highspeed, services remains seriously inadequate…"); George Gilder, "Broadband's Narrow Minds," *The Wall Street Journal*, February 19, 2003 ("Just three years ago, the U.S. was overwhelmingly dominant in communications technology and deployment … Today, the U.S. is falling precipitously behind.").

debate is over arcane topics such as defining broadband (information service vs. telecommunications service) and who gets access to the telephone lines and cables reaching consumers. In practical terms, the issues pertain to the regulatory conditions placed on providing access to existing broadband infrastructure and promoting an environment in which existing technologies – principally, cable modem and digital subscriber line (DSL) – can compete effectively against each other.

The premise of much policy pronouncement in the U.S. is that the greater is the geographic coverage of broadband availability, the greater broadband adoption and (of course) the greater the resulting social welfare. Needless to say, the determinants of broadband adoption are more complex and include both demand and supply factors. On the demand side, many have argued that there will be no mass-market adoption of broadband, regardless of availability, until sufficiently attractive broadband applications emerge that make the additional cost of broadband access worthwhile (known as the "killer app" theory).[4] Naturally, others argue that there will be no substantial investment in developing and deploying a "killer app" until broadband availability is sufficiently widespread that the investment in applications is worthwhile.[5] On the supply side, substantial investment is necessary either to upgrade existing cable infrastructure from a television-centric technology to one that permits two-way digital broadband Internet access,[6] or to extend fiber optic cable into the neighborhoods in order to expand the geographic footprint of availability of DSL service.[7] The expected return on such investments depends not only on investors' expectations about forthcoming demand, but also on the extent of competition that is likely to arise. Finally, consideration must be given to the obligations of incumbents to unbundle their broadband infrastructure and provide pieces of it at regulated rates that may not be compensatory for the risks and costs assumed by the investors.

Unbundling obligations imposed on one technology affect the desirability of investment in all broadband technologies. Unbundling obligations imposed on incumbent telecommunications providers' DSL facilities decrease those carriers' incentives to invest in DSL build-outs and, thereby, affect the cable and wireless providers' desire to invest in broadband technologies. On the one hand, it may decrease the urgency of cable providers to expand their geographic presence and gain any first-mover advantage won thereby. On the other hand, it might encourage cable deployment in areas that would otherwise be

---

[4] Howell (2002) observes that content availability is the primary factor "stimulating demand for cable products, [and] thereby creating competitive pressure on DSL offerings. While local loop unbundling posits faster response to this competitive pressure...the evidence implies that application cost-benefit tradeoffs are the primary drivers of broadband uptake."

[5] See, for example, William E. Kennard, *Consumer Choice Through Competition*, Remarks at the National Association of Telecommunications Officers and Advisors 19th Annual Conference, September 17, 1999 ("Our challenge today is to make broadband happen ... The demand is there. Americans want it. If we can get it there, it will open up a whole new horizon for electronic commerce in this country."); and TechNet, "TechNet CEOs Call for National Broadband Policy," January 15, 2002 (John Chambers, President and CEO Cisco Systems and Co-Founder of TechNet contents that "[i]n order to stay competitive, educate the workforce and increase productivity, the United States must have ubiquitous broadband.").

[6] Figures from the NCTA indicate that over the past five years, the cable industry has invested nearly $50 billion "to upgrade more than three-quarters of a million miles of plant with fiber optics." National Cable & Telecommunications Association, "Remarks of Robert Sachs, NCTA President and CEO at Cable 2001, 'We're Making Broadband Happen,'" Press Release, June 11, 2001.

[7] SBC, for example, announced in early 2000 an ambitious $6 billion investment, dubbed "Project Pronto," that seeks build out its fiber plant and, accordingly, extend the reach of its DSL services. Computer Science and Telecommunications Board (2002), pp. 259-263.

uneconomical given the prior presence of DSL. In either case, however, where cable providers have no competition from DSL technology (or vice versa), one would expect a decrease in the sole provider's incentives to compete vigorously for customers. Hence, one might anticipate supply factors that would drive broadband adoption to include the extent to which there is direct competition across broadband technologies (intermodal competition) at any level of overall availability of broadband services.

It might be predicted, a priori, that an empirical examination of the effect of availability on penetration would show a positive relationship, namely, more availability, more penetration. However, our discussion suggests that once the effect of competition is taken into account, the relationship may not be so straightforward. In fact, it is not clear which factor would have a more powerful impact on broadband adoption: the raw availability of broadband services, or the extent of competition across broadband platforms. This theoretical ambiguity can have unexpected effects on the relationship between total availability and penetration. Consider, for example, two hypothetical states: in one state, half the residents have DSL available, but not cable, and half have cable, but not DSL; in the other state, half the residents have no broadband available, but the other half have both platforms available. In the first state, the provider would price to maximize profit given the lack (in the short run) of an alternative provider in the state. An optimal price in the first state, given the absence of a competitor is most likely to limit the fraction of customers purchasing the service. In the second state, however, the normal mechanisms of competition between the platforms would drive price, advertising, quality, and other factors to encourage penetration, within the area in which the platforms are available. Hence, the first state might have a higher or lower overall level of penetration than the second.

II. Objective

The objective of this study is to offer empirical evidence directly relevant to the policy debate. To date there has been little research responsive to the issues raised and the limited research that has been conducted is only partially relevant. This could be attributed to the sheer novelty of the issue. Likewise, it might be due, in part, to the difficulty in modeling the relationships between competition, availability, and broadband adoption; namely, the nature of broadband availability and competition are clearly endogenous to a model of broadband adoption. In this study, we examine the influence of availability, competition, and demographics on the adoption of broadband technology in the U.S. The focus of our study is on the effect of intermodal competition on broadband penetration, relative to the effect of simple broadband availability. Our efforts to model this relationship should be viewed as a first step to bringing empirical evidence to the table. We take measures to control for the endogenous effects of competition and availability on adoption, but we admit, this effort is imperfect. Nevertheless, we believe the strength of the results and their intuitive appeal are compelling.

To motivate the empirical analysis, consider the following stylized representation of the broadband penetration. For purposes of this exposition, we model penetration very simply. That is, we assume that penetration in areas in which broadband is available is a function of whether there is or is not intermodal competition (we assume that if there is no availability of broadband service, broadband penetration in such areas is zero). Let P(O) represent the percent of broadband penetration in geographic regions in which there is "overlap," by which we mean head-to-head intermodal competition between cable modem and DSL platforms; that is, P(O) is the number of broadband subscribers in the geographic regions in a state in which both platforms are available, divided by the population of those geographic regions. Let G(O) represent the fraction of the state's population who reside in the geographic regions in which there is overlap availability (i.e., in which both DSL and cable broadband services are available). Similarly, Let P(E) represent the percent of broadband penetration in geographic regions in which one or the other platform is available, but not both ("exclusive" availability); that is, P(E) is the number of broadband subscribers in geographic regions in a state in which one or the other platform is available, but not both, divided by the number of residents in such regions. Let G(E) represent the fraction of the state's population who reside in the geographic regions in which there is exclusive availability of one or the other platform. In this notation, total broadband availability in the state, TA, is G(O) + G(E). With this notation, total broadband penetration in a state (TP) would be:

$$TP = P(O)G(O) + P(E)G(E)$$

We define "total penetration" in a state to be the number of subscribers divided by the state's population; that is, penetration measures actual subscribership, while availability reflects the extent to which customers could subscribe.

One could then ask, as a simple mathematical relationship, how does TP vary with the availability measures, G(O) and G(E)? The relationship must recognize that, empirically, there may be a positive or negative underlying relationship between G(O) and G(E). Both of these variables are determined by factors that affect the attractiveness of the market for broadband providers, including expected intensity of demand and expected costs of service. Suppressing these factors, in reduced form:

$$\partial TP/\partial G(O) = P(O) + P(E)\partial G(E)/\partial G(O)$$

and

$$\partial TP/\partial G(E) = P(E) + P(O)\partial G(O)/\partial G(E)$$

Hence, depending on the signs of the cross partials between G(E) and G(O), penetration may increase or decrease with G(E) and G(O), when $P(O) \neq P(E)$. If $P(O) = P(E)$ (i.e., if intermodal competition has no effect on broadband penetration), then total penetration is simply a function of total availability, and one would expect, unambiguously, a positive relationship between penetration and availability. However, if G(O) and G(E) are negatively correlated across states, then the sign of the effects of overlap availability and exclusive availability on total penetration would depend on the magnitude of the correlation and the relationship between P(E) and P(O). If, for example, there were no effect of competition on penetration ($P(E) = P(O)$), then an increase in overlap availability would be associated with an increase in total penetration as long as overlap penetration were positively correlated with total penetration (i.e., $\partial G(E)/\partial G(O) > -1$); and an increase in exclusive availability would be associated with an increase in total penetration as long as exclusive availability were positively correlated with total penetration (i.e., $\partial G(O)/\partial/G(E) > -1$). Our primary interest is in determining if intermodal competition (overlap availability) has a positive effect on penetration. Hence, we are interested in determining whether P(O) > P(E).

We model empirically the effects of both overlap availability (intermodal competition) and exclusive availability on broadband penetration, controlling for demand variables, cost variables, and other factors. In doing so, we adopt a weighted logit model that employs a single cross-section of state level (grouped) data. The model controls for demand and cost factors that influence the level of broadband adoption in a state, such as education, Internet access, line density and average loop length. The focus of primary importance in this paper is the influence of the overall availability of DSL and cable modem services in a state, and the overlap of this availability to both cable modem and DSL technologies. We cannot directly measure P(O) and P(E). However, we can address the question of interest by estimating the following linear model:

$$TP = \alpha + \beta(G(O), G(E), \text{demand controls, cost controls}).$$

In this model, the coefficient on G(O) estimates P(O), and the coefficient on G(E) estimates P(E).

The econometric results indicate that, in fact, P(O) appears to exceed P(E), controlling for demand and cost drivers. That is, we conclude that after controlling for the demand and cost influences on adoption, intermodal competition drives increased penetration in a state. In fact, while total broadband penetration in a state increases with total broadband availability in the state, the effect of total availability disappears when intermodal competition is controlled for. It appears, then, that broadband availability in a state is driven by intermodal competition and the demand and cost factors, but not by raw availability of broadband services, given those other factors. The independent effect of direct, intermodal competition is statistically significant and of substantial magnitude.

The organization of this paper is as follows. Section III summarizes previous studies that have addressed Internet and broadband deployment and adoption. Section IV provides a description of our econometric model, the variables employed in the model, and data sources. In Section V, we present our results, and in Section VI, we conclude with a discussion of the policy implications of our findings.

III. Previous Studies That Have Addressed Broadband Deployment and Adoption

To our knowledge, the first econometric study to examine broadband adoption is Madden, Savage, and Simpson (1996) (MSS) who examine survey data on a sample of Australian households. The survey asked participants to describe their preferences for a yet-to-be deployed broadband service. The authors discover a variety of demographic characteristics that influence an individual's decision to adopt broadband. They find, among other things, that individuals without a secondary degree express less interest in broadband; individuals residing in households with at least one member originating from Europe or Asia express greater interest; and age matters – respondents expressed less interest in broadband if one or more people in the household was 65 years or older. Surprisingly, a negative but statistically non-significant coefficient is found on the variable controlling for households with Internet access.

Rappoport, Kridel and Taylor (2002) (RKT) estimate the demand for broadband but within a slightly different context than MSS. RKT rely on survey data comprised of demographic information on a sample of households from across 10 U.S. cities. The households in the sample all subscribe to Internet access, but their choice of access is split between narrow-band (dial-up) and broadband. The authors separate the data into two groups, those who have chosen narrowband access and those that have chosen broadband access. The focus of their paper is, accordingly, on the characteristics that distinguish broadband users from narrowband users, rather than on what distinguishes Internet users from non-users (as does MSS). They find, using a probit model, that income and household size affect positively the likelihood of selecting broadband over dial-up. A somewhat surprising result of their study is the correlation between age and broadband subscribership: older users, they estimate, are more likely than younger users to choose broadband over dialup.

Another recent study that considers the demographic influences on adoption is the U.S Department of Commerce (2002). The study explores the same set of issues as MSS, except that it employs survey data on a sample of U.S. respondents and considers actual adoption patterns (as opposed to anticipated adoption, in the case of the Australian study). Not surprisingly, the Commerce study reached conclusions similar to those of MSS. In particular, the Commerce study finds that among the adult population, a smaller proportion of people aged 50 or older use the Internet than do those in younger adult age groups. Likewise, ethnic groups are found to have different levels of use; 50 percent of those surveyed in the White and Asian ethnic groups use the Internet, compared to 29 and 24 percent in the Black and Hispanic groups. The study finds a stark difference in Internet use by education group. For instance, only 13 percent of those surveyed with less than a high school education used the Internet, while 75 percent of those surveyed with a Bachelors degree or higher used the Internet.

While each of the above studies provides us with valuable information about who is more or less likely to be using the Internet or adopt broadband Internet access, they do not tell us whether or how the outcomes are affected by availability or how competition affects uptake. In sum, the Commerce study, MSS and RKT provide insight as to the demand characteristics associated with adoption, but do not tell us how cost and market structure factors affect adoption.

Gabel and Kwan (2000) address this deficiency in the literature by examining the cost and demand influences on the availability of residential broadband Internet access. The authors compile a detailed data set comprised of observations at the wire center level across the U.S. Using a stratified sampling method, they select a sample of nearly 300 observations. They employ a logit model to estimate demand and supply factors that influence the probability that DSL or cable modem are available in a wire center service area. Of the demand variables that are statistically significant, the authors find that where adults between 30 and 34 years in age are in greater number, there is a greater probability of deployment. Likewise, median income has a significantly positive influence on deployment. The cost variables they consider that are found to be statistically significant are teledensity (i.e., the number of telephone lines per square mile) and the estimated cost of transporting data traffic to the Internet backbone. The coefficients on both variables are negative, as anticipated. An interesting conclusion of their study is that none of the coefficients on the RBOC dummy variables in their regression are statistically significant. Based on this finding, they conclude that "residential customers in areas served by Regional Bell Operating Companies, all else equal, are equally likely to have high-speed access to the Internet as customers served by Independent telephone companies."[8]

In a report that perhaps most directly addresses the areas of interest in our study, the Organization for Economic Cooperation and Development (2001) (OECD) examined the adoption of broadband technology across the 30 OECD member countries. Unlike OECD (2001), however, our study takes an econometric approach rather than a survey approach, and we examine the variations in broadband adoption across U.S. states, rather than the variations across countries. The OECD report found that "[o]ne of the key ingredients in why some countries are forging ahead, is whether there is competition between different networks and networks with different technologies," and that "[t]here is a significant correlation between the growth of cable modems and DSL services."[9]

Countries where cable and DSL compete (rather than being jointly owned by the same entity) generally have a stronger broadband market. In Belgium, for example, Telenet launched its cable Internet access service as early as 1997. The existence of cable as a competitive alternative had a significant impact on the incumbent carrier's efforts to market DSL. As a result, Belgium is among the OECD countries with greater broadband penetration. Canada, ranked second in the OECD in terms of broadband penetration, is another case in point. Cable networks in Canada were among the first to introduce cable modem services, dating as far back as 1996. This early launch of broadband access with cable technology almost immediately spurred the reaction of telecommunications carriers. In November 1996, Sasktel became the first carrier in the OECD to offer broadband access using DSL technology. By June 2001, broadband penetration in Canada was close to five subscribers per 100 inhabitants, twice as much as the U.S. penetration rate, which is 2.25 per 100 inhabitants according to the same OECD study.

Another key finding in the OECD report is that both competition between technologies (intermodal competition) and among the facilities-based providers of the same technology (intramodal facilities-based competition) are important factors in the development of broadband markets. The report points to the case of South Korea, where the penetration rate of 9.2 subscribers per 100 inhabitants exceeds by far that of all other OECD countries. In South Korea, there are multiple companies competing with their own infrastructure, be it cable or DSL. The OECD study concludes that "[t]he most fundamental policy available to OECD governments to boost broadband access is infrastructure competition."[10]

The existing research provides a helpful starting point in developing our study. However, the focus of our investigation is different from and complementary to the existing research. Our study investigates the demand and cost influences on broadband adoption, as do the existing studies. But the primary focus of our study is an issue that has not been dealt with previously in an econometric fashion; namely, we seek to estimate the impact on adoption of variation in the competitiveness of availability, from instances of no availability, to single platform availability, to multiple platform availability. From this we hope to discover how and to what extent intermodal competition influences broadband uptake in the U.S. This is a

[8] Gabel and Kwan (2000), p. 1.
[9] OECD (2001), p. 9.
[10] OECD (2001), p. 4.

difficult issue to investigate empirically and it entails certain shortcomings that we describe below. Our study is motivated by the absence of empirical literature on this important topic and the urgency with which a better understanding is needed to properly devise policy in this area.

IV. The Empirical Model

To determine the impact of demographic and market structure characteristics on broadband adoption, this paper examines the following reduced form relationship:

PENETRATION = *f*(EDUCATION, INTERNET, TELEDENSITY, LENGTH, PLOOP, AVAILABILITY, OVERLAP, SKEW, BELLSOUTH, VERIZON, QWEST)

Where each variable is measured at the state level and defined as:

PENETRATION	=	The number of broadband subscribers in the state, weighted by population, as of December 2000;
EDUCATION	=	The percentage of the population in the state, 25 or over with a Bachelors degree or higher, as of December 1990;
INTERNET	=	The percentage of households in the state with Internet access, as of October 2000);
TELEDENSITY	=	The number of incumbent local exchange switched access lines in the state, per square mile, as of December 2000;
LENGTH	=	The average length of an incumbent local exchange switched access line in the state, as of December 1998;
PLOOP	=	The regulated price for a Regional Bell Operating Carrier (RBOC) unbundled network element (UNE) loop in Zone 1 (urban areas) in the state, as of May 2001;
AVAILABILITY	=	The percentage of the population in the state residing in a city where cable modem or DSL are deployed, as of April 2000;
OVERLAP	=	The percentage of the population in the state residing in a city where cable modem and DSL are deployed, as of April 2000;
SKEW	=	The percentage by which cable modem penetration exceeds DSL penetration in the state, as of June 2002;
BELLSOUTH	=	Binary (1/0) variable indicating states where BellSouth is the RBOC (1);
VERIZON	=	Dummy (1/0) variable indicating states where Verizon is the RBOC (1); and
QWEST	=	Dummy (1/0) variable indicating states where Qwest is the RBOC (1).

Descriptive statistics and data sources are provided in Table 1.

Table 1: Variable Names, Descriptive Statistics, and Sources

Variable	Mean	Standard Deviation	Source
PENETRATION	150,353	228,839	a
POPULATION	6,101,113	6,386,824	b
EDUCATION	0.1298	0.0245	b
INTERNET	0.4999	0.0646	c
TELEDENSITY	108	168	a, b
LENGTH	14,461	3,112	a
PLOOP	13.51	4.73	d
AVAILABILITY	0.2411	0.1201	e
EXCLUSIVE	0.1346	0.0881	e
OVERLAP	0.1064	0.0989	e
SKEW	0.3688	0.2165	a
BELLSOUTH	0.1957	0.4011	f
VERIZON	0.2174	0.4170	f
QWEST	0.2826	0.4552	f
SBC	0.2826	0.4552	f

Sources:

(a) Federal Communications Commission, Industry Analysis and Technology Division, *High Speed Services for Internet Access* (various reports).

(b) U.S. Census Bureau, various reports.

(c) U.S. Department of Commerce, National Telecommunications & Information Administration, *Falling Through the Net: Toward Digital Inclusion*, October 2000, Table I-B.

(d) Billy Jack Gregg, *A Survey of Unbundled Network Element Prices in the United States*, April 2001.

(e) Rural Utilities Service & National Telecommunications & Information Administration, Joint Project, *Advanced Telecommunications in Rural America: The Challenge of Bringing Broadband Service to All America*, Appendix A and B, April 2000.

(f) Company websites.

The dependant variable in our analysis, PENETRATION, consists of total broadband subscribership in a state. Our source for data on PENETRATION is the FCC's biannual report on broadband adoption, *High Speed Services for Internet Access*. The report includes, among other things, two measures of broadband adoption: (1) total subscribership, and (2) small business and residential subscribership. The results of our study are effectively unchanged by our choice of total subscribership as opposed to total small business and residential subscribership. There are, of course, other possible dependent variables that are interesting to consider, most obviously the price of broadband within a state. We do not consider price in our study due to the absence of available data.[11]

The independent variables in the above equation can be classified into four categories:

[11] Moreover, the extent of cross-jurisdiction variation in prices is not clear. The "rack rate" price observed from providers' websites, for instance, may differ from their actual region/city-specific prices, depending on the promotional offerings available in that market (such as a waiver of installation and/or equipment charges, a limited introductory rate, a discount for purchasing a bundled offering, etc.). However, quantifying this region/city specific variation information (assuming it exists) can be difficult. Hausman, Sidak and Singer (2001) did not observe significant cross-jurisdiction variation in the price of cable modem service in their study.

i. Demand Controls. The variable INTERNET is included to control for the hypothesized complementary relationship of Internet access to broadband adoption. Therefore, a positive estimated coefficient is expected.[12] We include the variable EDUCATION to control for the hypothesized positive influence that education has on demand.[13] This variable may also partially control for the influence of income on demand, which one would also expect to be positive. A preliminary regression was run controlling for the influence of income. It was found to be statistically insignificant and the results of an F-test indicated that we should leave the income variable out of the regression.

ii. Cost Controls. TELEDENSITY is included to control for the cost of deploying broadband technology. The decision to deploy broadband should be less costly in those states where there are more switched access lines per square mile. Therefore, a positive coefficient on TELEDENSITY is anticipated.[14] Likewise, LENGTH affects the cost of deploying broadband technology and, in particular, DSL technology. The speed of DSL service diminishes with the distance from a customer's home to the central office, and typically not available to customers with loops greater than three miles in length. Measures to extend the reach of DSL to these customers are technically feasible but entail additional cost.[15] The coefficient on LENGTH should be negative. Finally, PLOOP is strictly controlled by state regulators and is a major input cost to competitive local exchange carriers (CLECs) offering broadband services via ILEC facilities. To some extent we might also expect PLOOP to proxy for the costs to the ILEC, though this relationship is likely to be noisy. Both possibilities suggest a negative coefficient on PLOOP.

iii. RBOC Controls. The dummy variables BELLSOUTH, VERIZON, QWEST, and SBC are intended to control for the behavior of the primary DSL provider (as proxied by the particular RBOC present in the state) across states that is not controlled for by the demand and cost variables. SBC is the excluded RBOC in our model. To the extent that, for instance, one RBOC pursues a more aggressive broadband strategy than another RBOC, all things equal, this would be reflected in a positive coefficient on the RBOC. There are, however, many possible region-wide factors that might influence the sign of the coefficient on the dummy variables. Accordingly, we do not have a preconceived notion as to the anticipated coefficient on these variables.

iv. Market Structure Controls. Consistent with our previous discussion in Section 2, an increase in AVAILABILITY ought to have a positive impact on broadband adoption; in those states where broadband services are more widely available, we should observe greater adoption. The data on AVAILABILITY, however, allow for disaggregation of this variable

[12] Madden, Savage, and Simpson (1996) include a control for internet access in their pioneering study of survey data on the demographic criteria that influence household subscription to broadband in Australia.

[13] This hypothesized relationship is made by Gabel and Kwan (2000) and Madden, Savage, and Simpson (1996). Both studies uncover the anticipated coefficient, however, only the latter study finds a statistically significant relationship.

[14] A more accurate control for line density would measure both the volume and the distribution of lines within the state. For instance, consider two states with the same square mileage and the same number of telephone lines. In one state, the lines are scattered uniformly throughout the state, while in the other state the lines are concentrated within a single city block. Our measure of TELEDENSITY will be the same despite the differences in density attributed to the distribution of lines in the two states. Ideally, we would like to be able to disaggregate TELEDENSITY (as well as other controls) across the overlap and exclusive territories, and our inability to do so clearly decreases the effectiveness of the controls.

[15] For instance, SBC announced a $6 billion DSL loop renovation program known as "Project Pronto" that will extend the reach of DSL services to customer locations where it was previously unavailable. Computer Science and Telecommunications Board (2002), pp. 259-263.

into two components: EXCLUSIVE, the percentage of state population with access to one platform and only one platform, cable modem *or* DSL; and OVERLAP, the percentage of state population with access to two platforms, cable modem *and* DSL. The data for AVAILABILITY and its components are based on observation of cable modem or DSL deployment at the city level, and in some instances at the county level.[16]

We anticipate the coefficients on the variables OVERLAP and EXCLUSIVE to be nonnegative, consistent with our prediction of the sign of the coefficient on AVAILABILITY. The objective of this analysis, however, will be to examine whether overlap has an incremental positive effect, controlling for the other factors that we expect to influence penetration.

The variable SKEW is distinct from AVAILABILITY (as well as EXCLUSIVE and OVERLAP) in that the former does not consider how broadband is *available*, but instead controls for the extent to which *actual penetration* is dominated by a single platform. In every state in our sample, except California, Kentucky, and Utah, the share of cable modem exceeds that of DSL. Given that we are examining penetration levels at an early stage in the development of broadband, we expect that cable modem's first-mover advantage will be reflected in a positive coefficient on SKEW.

V. The Empirical Results

Our data consist of observations on 46 U.S. states. Observations on the variable SKEW were not available for Delaware, Hawaii, Rhode Island, and Wyoming. The District of Columbia was excluded because it was an outlier for many of the variables of interest. Including the District of Columbia in the sample did not affect significantly the coefficients or their signs but it did diminish the model's adjusted R-squared. To estimate the determinants of broadband adoption we employ a weighted least-squares logit regression. Because the dependent variable is constructed from grouped data (total broadband subscription in a state, divided by the state population) with unequal size groups, the OLS assumption of uniform variance is violated; that is, given populations of varying size over the states that make up the sample, heteroskedasticity will be present across the disturbances. The solution in this case is to model variation in broadband adoption with weighted least-squares estimates in a grouped logit equation.[17]

To understand the influence of availability on penetration, we present the regression results of two model specifications. Our first specification examines the influence of overall availability on adoption, and our second specification examines a disaggregated measure of availability. In particular, we disaggregate total availability (AVAILABILITY), into what we previously defined as G(O) and G(E). The empirical notation for these variables is OVERLAP and EXCLUSIVE, respectively. The grouped logit regression results of the aggregate (AVAILABILITY) model are presented in Table 2, and the regression results of the disaggregated (OVERLAP + EXCLUSIVE) model are presented in Table 3.

[16] As Gabel and Kwan (2000) indicate, broadband deployment within a city may not be ubiquitous. Our data do not permit us to determine the extent to which this is or is not the case.

[17] The models below are run with the -glogit- command of Stata 7.0. Glogit estimates weighted least squares, accounting for different sized denominators (state populations) and, in principle, different variances across states.

A. The Restricted Model

Overall, the model with AVAILABILITY performs well, explaining 71 percent of the variance in broadband adoption. Moreover, many of the coefficients of interest are statistically significant (at the 10 percent level or better) and all achieve their expected signs (of the coefficients for which we predict an effect). The magnitude of these effects can be seen by simulating expected levels of broadband adoption, given specified changes in the values of the independent variables. These results are presented in Table 2, column (3). The range of the specified change in the independent variable is assumed to be one standard deviation from the mean of the independent variable, except in the case of a dichotomous variable where we assume a change between 0 and 1. Given their intuitive interpretation, we will refer to column (3) results in our discussion below.

Both of the demand controls, INTERNET and EDUCATION, have a positive effect on penetration, consistent with our hypothesis. However, only EDUCATION is statistically significant. Moreover, EDUCATION is found to have a profound impact on penetration; the model predicts that a one standard deviation increase in EDUCATION in a state increases the per capita penetration level by 96 percentage points. The positive effect of education is consistent with the results of the U.S. Department of Commerce (2002), discussed earlier. The size of the coefficient may reflect the fact (we believe) that education is capturing demand, and demand is a powerful determinant of both entry and penetration. Likewise, each of the cost control variables – LENGTH, TELEDENSITY, and PLOOP – achieve their expected sign. TELEDENSITY is statistically significant (at the 5 percent level). The coefficient on TELEDENSITY suggests that a one standard deviation increase in the density of switched access lines results in a 0.0016 percentage point increase in per capita penetration. It is interesting to note that, while the coefficient on PLOOP has the expected sign, it is not statistically significant. This might be considered surprising in light of the perceived importance of UNE pricing on promoting penetration. It is possible (though speculative) that this result suggests that CLECs deploying DSL using UNEs have little effect on overall broadband penetration. It may also be that CLECs, rationally, make entry and investment decisions not on just the basis of current UNE prices but on the basis of expected future prices. Given the substantial uncertainty that existed at the time, and still exists, regarding future UNE prices, it is not surprising to find weak correlation between contemporaneous UNE prices and CLEC decisions. Notice that while we do not control for the extent to which provisioning of UNEs varies across states, which might be correlated with PLOOP, the influence of RBOC-specific variation is captured by the RBOC dummy variables.

LENGTH is marginally significant at the 15 percent level, slightly below the standard threshold of statistical significance (10 percent or better). The coefficient on LENGTH suggests that a one standard deviation increase in the average length of ILEC loop reduces penetration by 0.00005 percent. The coefficient on SKEW, which controls for the percentage cable modem penetration exceeds (or is less than) the penetration of DSL, is positive and statistically significant. We interpret the coefficient on SKEW as identifying the advantage cable modem derives from a first-mover advantage into the broadband marketplace. A one standard deviation increase in the percentage cable modem penetration exceeds DSL is estimated to result in a 1.2 percentage point increase in per capita penetration.

Table 2: Weighted Least-Squares Logit Regression
Dependent Variable: PENETRATION
Weight: POPULATION
Specification: Restricted Model

Variable	(1) Coefficient Estimates (SE)	(2) Odds Ratio	(3) Net Change in the expected per capita adoption of broadband
Intercept	-5.523* (0.581)	--	--
INTERNET	1.163 (1.088)	3.2009	5.018%
EDUCATION	7.677* (2.76)	2157.745	95.73%
TELEDENSITY	0.007* (0.00028)	1.0007	0.0016%
LENGTH	-0.00002 (0.00001)	0.99998	-0.00005%
PLOOP	-0.012 (0.012)	0.9878	-0.029%
AVAILABILITY	1.176* (0.396)	3.242	5.107%
SKEW	0.396* (0.185)	1.486	1.155%
BELLSOUTH	0.345* (0.135)	1.413	.982%
VERIZON	0.033 (0.111)	0.834	-0.399
QWEST	-5.523 (0.137)	1.033	0.08%
Observations	46		
F statistic	12.05		
R-Squared	0.775		
Adjusted R-Squared	0.7107		

Notes: The entries in column (1) are weighted least-squares logit estimates for grouped data (standard errors in parenthesis). Columns (2) and (3) are based on the exponential linear predictions generated by the -or- routine in Stata 7.0, and they are calculated assuming all other variables are set at their mean values. The net change in the expected per capita adoption of broadband is a simulated probability calculated by adjusting the independent variables by one standard deviation from the mean value for the continuous variables and between 0 and 1 for the dichotomous variables.

* Significant at the 5% level; and ** Significant at the 10% level.

The coefficient of primary interest in the restricted regression is that on AVAILABILITY. As expected, the coefficient on this variable is positive, it is statistically significant, and its magnitude is relatively large. We estimate that a one standard deviation increase in AVAILABILITY results in over a 5 percentage point increase in the per capita penetration level. Importantly, however, the regression omits disaggregated information on the nature of AVAILABILITY in state, the impact of which we investigate in our regression on the unrestricted model.

Before describing the regression results of the unrestricted model, we should first comment on the potential endogeneity of AVAILABILITY and SKEW. In particular, it is possible that even after including the demand and cost controls there might be unobserved effects that are correlated with cost and/or demand and AVAILABILITY and SKEW. We do not have a definitive solution to this problem; there are no instruments of which we are aware that can be employed to control for the potential endogeneity of AVAILABILITY and SKEW. We can, however, assess the effectiveness of the demand and cost controls that are included in the model. The more effective these controls are, the more reassurance we can have that our model controls for the major demand and cost drivers, and the less distortion the unobserved variables should have on AVAILABILITY and SKEW. To examine the impact of our cost and demand variables on the fit and coefficient estimates on AVAILABILITY and SKEW, we run the restricted model omitting the demand and cost controls. The results of this regression are presented in Appendix A. They suggest that the demand and cost controls employed in the model have an important impact on regression results. The adjusted R-squared is 27 percent, 47 percentage points less than the adjusted R-squared that results when including the demand and cost controls. Moreover, the coefficient on SKEW changes sign and loses significance. AVAILABILITY, in contrast, retains its sign and statistical significance.

B. The Unrestricted Model

In this model, we disaggregate AVAILABILITY into its component parts, EXCLUSIVE and OVERLAP. We conjecture that the impact of broadband availability on penetration in locations where both DSL and cable modem are present (OVERLAP) will differ from its impact on penetration in locations where only a single platform is available (EXCLUSIVE). As with the restricted model, the unrestricted model performs well, achieving a slightly higher adjusted R-squared of 74 percent, compared to the restricted model. All of the coefficients achieve their expected sign (of the coefficients for which we predict an effect) and most of the coefficients of interest achieve statistical significance at the standard levels (10 percent or better). As in our presentation of the restricted model, we estimate the magnitude of the effects of each coefficient by simulating expected levels of broadband adoption, given specified changes in the values of the independent variables. These results are presented in Table 3, column 3. Again, given their intuitive interpretation, we will refer to column (3) results in our discussion below.

Table 3: Weighted Least-Squares Logit Regression
Dependent Variable: PENETRATION
Weight: POPULATION
Specification: Unrestricted Model

Variable	(1) Coefficient Estimates (SE)	(2) Odds Ratio	(3) Net Change in the expected per capita adoption of broadband
Intercept	-6.048* (0.612)	--	--
INTERNET	2.407** (1.204)	11.096	19.43%
EDUCATION	8.23* (2.65)	3754.17	96.5%
TELEDENSITY	0.0005** (0.00028)	1.0005	0.0013%
LENGTH	-0.00003* (0.00001)	0.9999	-0.0001%
PLOOP	-0.006 (0.012)	0.9936	-0.02%
EXCLUSIVE	0.0089 (0.893)	1.009	0.02%
OVERLAP	1.357* (0.389)	3.884	6.47%
SKEW	0.419* (0.178)	1.52	1.23%
BELLSOUTH	0.399* (0.132)	1.49	1.17%
VERIZON	-0.256* (0.112)	0.774	-0.546
QWEST	-0.061 (0.139)	0.941	-0.142%
Observations	46		
F statistic	12.35		
R-Squared	0.7998		
Adjusted R-Squared	0.735		

Notes: The entries in column (1) are weighted least-squares logit estimates for grouped data (standard errors in parenthesis). Columns (2) and (3) are based on the exponential linear predictions generated by the -or- routine in Stata 7.0, and they are calculated assuming all other variables are set at their mean values. The net change in the expected per capita adoption of broadband is a simulated probability calculated by adjusting the independent variables by one standard deviation from the mean value for the continuous variables and between 0 and 1 for the dichotomous variables.

* Significant at the 5% level; and ** Significant at the 10% level.

There is little change in the coefficient estimates on the variables appearing in both equations. Therefore, we will not repeat these findings. The coefficient estimates of interest in this regression are those on the availability variables, OVERLAP and EXCLUSIVE. Both coefficient estimates are positive, as anticipated. However, they have substantially different effects on penetration. The coefficient on OVERLAP is quite large and statistically significant (at the 5 percent level); a one standard deviation increase in overlap availability is estimated to increase per capita penetration by 6.5 percentage points. In contrast, the magnitude of the coefficient estimate on EXCLUSIVE is small and is not statistically significant. In interpreting this result, it is informative to compare it to the equivalent regression excluding the demand and cost variables. The results of this regression are found in Appendix B. The adjusted R-squared is 28 percent, compared to 74 percent in the regression including demand and cost variables. However, the coefficient estimates on both OVERLAP and EXCLUSIVE are positive and statistically significant (at the 5 percent level). Both variables are estimated to have large effects on penetration, but the relative magnitude of their effect is the opposite of that found when we include demand and cost variables in the regression. A comparison of these two regressions suggests that after controlling for the influences of demand and cost, OVERLAP retains its magnitude and its statistical significance, while EXCLUSIVE falls to zero and loses its predictive power. An F-test confirms (at the 5 percent level) that the coefficients on these two variables are statistically different from each other. In fact, an F-test suggests that EXCLUSIVE is not needed in the regression. The regression results when EXCLUSIVE is dropped achieve a marginally higher adjusted R-squared and the coefficients estimates (and their t-statistics) are little changed from the output of the regression that retains EXCLUSIVE. In sum, the effect of exclusive availability is de minimus, if not zero, when controlling for the effect of overlap availability.

This result suggests that the demand and cost controls fully explain availability, and there is no incremental explanatory power of the raw availability measure, while intermodal competition (OVERLAP) retains significant additional explanatory effects. This interpretation is consistent with the results of an alternative specification we ran, in which we estimated the effects of overlap and total availability (OVERLAP and AVAILABILITY), rather than overlap and exclusive availability. We found that given overlap, the effect of total availability was not statistically significant given the presence of the other controls.

VI. Conclusion

Our results demonstrate a substantial and statistically significant positive impact on broadband adoption of intermodal (facilities-based) broadband competition, even controlling for broadband availability. These results are consistent with the observations made by OECD (2001), in its cross-national study discussed earlier in our paper. We believe that in providing econometric evidence of this effect, our results are unique and make a new contribution to the understanding of broadband adoption in the U.S. Our results are also not surprising, in that one would expect facilities-based competition to drive additional broadband adoption, over and above pure availability of the service. What is perhaps surprising is that controlling for demand and cost factors, and intermodal competition, exclusive availability (availability in areas without intermodal competition) has no additional predictive power for adoption.

As a policy matter, these results suggest that the focus in the current policy debate on ubiquitous access to broadband service may not be the most effective in driving adoption. Rather, policies that encourage facilities-based competition—and, therefore, policies that encourage facilities investment by all platform providers, are perhaps more important. Policies that discriminate across platforms, to the effect of encouraging investment in one (or more) but discouraging investment in another, will frustrate policy objectives to increase adoption.

We view our results as a first step at modeling and estimating the effects of competition on broadband adoption. It would be useful in further research on these issues, in our opinion, to seek to improve controls on the potential endogeneity of the supply factors. Ours is a reduced form model, and additional structure might add additional insights. In addition, our data on the demand and supply controls (education, teledensity, and so forth) are at the state level, and therefore do not specifically identify demand and supply factors within the overlap and exclusive territories. We are unaware at this time of more granular data, but to the extent such data can be developed, they would add useful refinement to the analysis.

REFERENCES

Computer Science and Telecommunications Board, National Research Council, *Broadband: Bringing Home the Bits*, National Academy Press (2002).

Federal Communications Commission, *High Speed Services for Internet Access: Status as of December 31, 2000*, August 2001.

Federal Communications Commission, *High Speed Services for Internet Access: Status as of June 30, 2002*, December 2002.

Ferguson, Charles H., "The U.S. Broadband Problem," The Brookings Institution, Policy Brief #105, July 2002.

Gabel, David, and Florence Kwan, "Accessibility of Broadband Telecommunications Services by Various Segments of the American Population," Paper Prepared for the Telecommunications Policy Research Conference, August 2000.

Gilder, George, "Broadband's Narrow Minds," *The Wall Street Journal*, February 19, 2003.

Gregg, Billy Jack, *A Survey of Unbundled Network Element Prices in the United States*, April 2001.

Hausman, Jerry A., Gregory Sidak, and Hal J. Singer, "Cable Modems and DSL: Broadband Internet Access for Residential Customers," American Economic Review, Vol. 9, No. 2, May 2001, pp. 302-307.

Howell, Browyn, "Broadband Uptake and Infrastructure Regulation: Evidence from the OECD Countries," New Zealand Institute for the Study of Competition and Regulation, Working Paper BH02/01, February 2002.

Kennard, William E., *Consumer Choice Through Competition*, Remarks at the National Association of Telecommunications Officers and Advisors 19th Annual Conference, September 17, 1999.

Landers, Jim, "Broadband believer: FCC's chair presses for big leap for U.S. high-speed Internet," *The Seattle Times*, October 21, 2002.

Madden, Gary, Scott Savage, and Michael Simpson, "Information Inequality and Broadband Network Access: An Analysis of Australian Household Survey Data," *Industry and Corporate Change*, Oxford University Press, 1996, pp. 1049-1056.

National Cable & Telecommunications Association, "Remarks of Robert Sachs, NCTA President and CEO at Cable 2001, 'We're Making Broadband Happen,'" Press Release, June 11, 2001.

Organization for Economic Cooperation and Development, *The Development of Broadband Access in OECD Countries*, October 29, 2001.

TechNet, "TechNet CEOs Call for National Broadband Policy," Press Release, January 15, 2002.

U.S. Department of Commerce, National Telecommunications & Information Administration, *Falling Through the Net: Toward Digital Inclusion*, October 2000.

U.S. Department of Commerce, and National Telecommunications & Information Administration, *A Nation Online: How Americans Are Expanding Their Use of the Internet*, February 2002.

Rappoport, Paul N., Donald J. Kridel, and Lester D. Taylor, "The Demand for Broadband: Access, Content, and the Value of Time," Working paper presented at the Southern Methodist University economics workshop, August 23, 2002.

Rural Utilities Service, and National Telecommunications & Information Administration, *Advanced Telecommunications in Rural America: The Challenge of Bringing Broadband Service to All America*, April 2000.

Wall Street Journal, "The Harris Poll: Online Population Continues to Expand but Just Barely," February 5, 2003.

APPENDIX A: WEIGHTED LEAST-SQUARES LOGIT REGRESSION

Dependent Variable: PENETRATION
Weight: POPULATION
Specification: Restricted Model, Excluding Demand and Cost Variables

Variable	(1) Coefficient Estimates (SE)
Intercept	-4.022* (0.204)
AVAILABILITY	1.32* (0.44)
SKEW	-0.243 (0.24)
BELLSOUTH	-0.039 (0.151)
VERIZON	0.214 (0.135)
QWEST	-0.007 (0.156)
Observations	46
F statistic	4.33
R-Squared	0.3513
Adjusted R-Squared	0.2702

* Significant at the 5% level
** Significant at the 10% level

APPENDIX B: WEIGHTED LEAST-SQUARES LOGIT REGRESSION

Dependent Variable: PENETRATION
Weight: POPULATION
Specification: Unrestricted Model, Excluding Demand and Cost Variables

Variable	(1) Coefficient Estimates (SE)
Intercept	-4.102* (0.215)
OVERLAP	1.04* (0.498)
EXCLUSIVE	2.022* (0.741)
SKEW	-0.182 (0.242)
BELLSOUTH	-0.042 (0.151)
VERIZON	0.253** (0.138)
QWEST	0.002 (0.155)
Observations	46
F statistic	3.87
R-Squared	0.3734
Adjusted R-Squared	0.277

* Significant at the 5% level
** Significant at the 10% level

PART III.
PUBLIC POLICY AND FUTURE DIFFUSION

Chapter 7

THE TRAGIC INEFFICIENCY OF THE M-ECPR*

Nicholas Economides
Stern School of Business
New York University

1. INTRODUCTION

In a recent article,[1] J. Gregory Sidak and Daniel F. Spulber ("SS") argue for the validity of the regulatory rule "M-ECPR," an adaptation of "ECPR." M-ECPR, or "Market Determined Efficient Component Pricing Rule" was proposed by Michael J. Doane, *et al.* (1996) at a regulatory proceeding in Hawaii in support of a filing by GTE Inc.[2] The original ECPR ("Efficient Component Pricing Rule") has been originally proposed by Robert Willig and William Baumol in the late 70s and early 80s.[3]

In this article, I argue against the validity of the M-ECPR (and consequently of the ECPR) as an appropriate regulatory rule in telecommunications. I will show that use of M-ECPR (and of ECPR) and, more generally of the *private* opportunity cost of the incumbent local exchange carriers ("ILECs") as a basis for setting prices for unbundled network

---

* The title derives from the original meaning of the word "tragic." In Ancient Greek tragedies, the hero, although valiant, is held captive by his own contradictions, apparently beyond his control, and this leads to his demise. Similarly, the M-ECPR and the ECPR are held hostage to original inefficient pricing, and this leads to their demise, since their application perpetuates inefficiency.

I thank Glenn Stover and participants in the American Enterprise Institute conference on "Costs and Pricing in Telecommunications" for their helpful comments.

[1] See Sidak and Spulber (1996).

[2] This paper was subsequently filed in many State regulatory proceedings on the implementation of the Telecommunications Act of 1996 as an attachment to the positions of GTE.

[3] Originally ECPR was proposed by Willig (1979). Baumol (1983) started its application in the railroad industry. It has also been used in the electricity industry. The adoption of the ECPR in the telecommunications sector in New Zealand (Privy Council decision) lead to significant long term damage in the sector, since entrants were unable to provide local service because of ECPR pricing of interconnection by the incumbent monopolist, Telecom New Zealand ("TCNZ"). In applying ECPR, TCNZ demands a very significant termination fee to terminate calls of competing carriers to its network, while refusing to pay any fees to other carriers for termination of calls to their networks when they originated from TCNZ's network. Economides, Lopomo and Woroch (1996) discuss ways to fix this interconnection problem through the adoption of *reciprocity* of termination charges.

elements ("UNEs") of the telecommunications network and more generally as a basis for interconnection pricing will lead to a perpetuation of inefficiencies. I also show that the use of M-ECPR is against the intent of the Telecommunications Act of 1996, which was to open *all* telecommunications markets to competition.

2. Definitions of the ECPR and the M-ECPR Pricing Rules

Consider a situation where a good or service AB is composed of two components, A and B, which, for simplicity, are assumed to be combined in a 1:1 ratio.[4] Assume further that firm 1 is originally a monopolist for goods A and B, and good A is required for the production of service AB; that is, A is a "bottleneck." For simplicity, let the unit cost of component A be c_A, and the cost of component B be c_B, while good AB is sold at price p_{AB}. Now suppose that a new firm (firm 2) enters the market for B and would like to produce service AB. The ECPR sets an "access fee," or price, for component A, to be paid by firm 2, equal to the pre-entry (monopoly) price of final good AB minus the cost of component B, that is, at $p_B = p_{AB} - c_B$. The ECPR would also set the same access fee for component A to any other entrant.

The M-ECPR sets the same fee as the ECPR, except for the case in which another competitor offers A at a lower price, resulting in price for AB $p'_{AB} < p_{AB}$. In that case, the M-ECPR sets the access fee for provision of component B to firm 2 at $p'_B = p'_{AB} - c_B$.

3. How Do the ECPR and the M-ECPR Differ?

In general, the ECPR and the M-ECPR coincide both in their logic and in the access fees they imply. The only difference of ECPR and M-ECPR can occur when the entrant, or a firm other than the incumbent monopolist, can provide component A at a lower price than the full monopoly price of the incumbent, resulting in a final price $p'_{AB} < p_{AB}$. Under these circumstances, the M-ECPR proposes an access fee for component A equal to $p'_B = p'_{AB} - c_B$. Thus, M-ECPR differs from ECPR *only if* the monopolized good is offered by a competitor (or self-provided by the entrant) at a price below that of the monopolist.

According to the authors of M-ECPR, this difference allows them to use the words "market determined" in describing their version of ECPR.[5] However, the use of the words "market determined" in this context is very misleading because the M-ECPR does not in any way try to imitate the prices that would exist in a *competitive* market. In fact, the M-ECPR is just a way to perpetuate the monopoly profits and historical productive inefficiencies of the incumbent, which would certainly be wiped out if a vigorous market existed, or if the M-ECPR imitated a competitive market. All the M-ECPR does is to define a monopoly price for an UNE. Like any profit-maximizing monopolist, the monopolist of component A using M-ECPR will not charge more than the price that may be quoted by an alternative provider. Thus, the M-ECPR corrects the original ECPR when it was resulting in prices above monopoly levels; the M-ECPR results in monopoly prices when an alternative provision may be available. Our criticisms below apply equally to the ECPR and M-ECPR.

---

[4] The criticisms of ECPR and M-ECPR also apply to more general cases of joint production.

[5] Doane *et al.* (1996) claim that the M-ECPR is "market-based," at page 4.

4. The Logic of the ECPR and the M-ECPR

The ECPR was designed to guard against productively inefficient entry. Since, under the ECPR, the entrant in the production of component B has to pay to the incumbent the full monopoly profits of the incumbent in the market for the composite product AB, the entrant will not survive unless the entrant is equally efficient, or more efficient than the incumbent in the production of component B. Thus, the ECPR and the M-ECPR are supposed to guard against productively inefficient entry. However, in attempting to achieve this task, the ECPR and the M-ECPR force consumers to pay a terrible price in terms of unrealized price decreases and lost consumers' surplus that would have been realized if these rules were not imposed. Application of the ECPR and the M-ECPR result in an allocative inefficiency. Moreover, often the loss in allocative efficiency that results from the use of the ECPR (or the M-ECPR) is much larger than any potential gains in productive efficiency from its use. Thus, in terms of total efficiency, use of the M-ECPR and the ECPR are detrimental to social welfare and to overall efficiency. Before establishing these propositions in the next sections, we briefly discuss the purported justification for ECPR and M-ECPR based on private opportunity cost.

5. Are the ECPR and the M-ECPR Cost-Based?

ECPR and M-ECPR are often presented as a resulting in access fees that are purely cost-based. Proponents of these rules say that the fee the ECPR and the M-ECPR set as the price that the entrant should pay for component A is just the actual cost of A plus the opportunity cost to the incumbent. Since profits are part of the opportunity cost of the incumbent, in this interpretation, the ECPR-based fee includes any original profits, normal and supernormal, as well as any cost inefficiencies of the incumbent. However, a fee that includes supernormal profits cannot possibly be thought of as containing only costs. Where is the mistake?

5.1 *Private* Opportunity Cost in Contrast with *Social* Opportunity Cost

The fallacy of the proponents of the ECPR and the M-ECPR lies in confusing *social* opportunity cost with *private* opportunity cost. Social opportunity cost of a resource reflects the present social cost of the resource and should be correctly included in a cost calculation. Private opportunity cost is the benefit or cost to a private party of a certain activity. Private opportunity cost differs in general from social opportunity cost, since private opportunity cost does not, in general, reflect the cost of resources to society, which social opportunity cost does.

An example will be helpful to understand the difference. Suppose that two companies, X and Y are competing for the business of customer C, which is worth \$C to each of them. Assume that X and Y are equally cost efficient in serving C. If customer C used to buy from X and now buys from Y, firm X's *private* opportunity cost is \$C. However, the *social* opportunity cost of the switch of customer C from X to Y is exactly zero, since society does not gain or lose from customer C's change of carrier. Essentially, since firm X's loss was firm

Y's gain, private opportunity costs and gains canceled each other, and the social cost of customer C's change of carrier is zero.

5.2 Prices Based on Social Opportunity Cost are Efficient But Prices Based on Private Opportunity Cost are Inefficient

Economic theory teaches that, to achieve allocative, productive and dynamic efficiency, *social* (rather than *private*) opportunity costs (and benefits) should guide pricing decisions. Private opportunity costs differ, in general, from social opportunity costs. In the pricing of access to monopolized bottleneck facilities, there is significant private benefit to the owner of a bottleneck facility in charging a high price for access to the facility. However, high prices of access would result in a significant social loss because they result in prices of final services that are higher than the efficient prices. Therefore, an incumbent monopolist should not be compensated for its (private) opportunity costs (that is, its lost profits) that result from entry of a rival.

6. The General Inefficiency of the ECPR and the M-ECPR

Economists use the word "efficient" to denote an outcome that optimizes an objective. In economics, typically "efficient" is meant to be "socially efficient," that is, maximizing total social welfare or total surplus. Does the ECPR or the M-ECPR, in general, maximize social surplus? Absolutely not! Although the rule has been debated for the last 18 years, neither its creators nor its present supporters have ever provided a *proof* that the use of either of these two rules maximizes social surplus, and thereby deserves to be called "efficient."

The lack of such proof over such a long period of time is telling. In fact Economides and White (1995, 1998) and Laffont and Tirole (1994) have proved the general *inefficiency* of the ECPR rule, and similar proofs apply to M-ECPR. The essence of the proof of Economides and White (1995, 1998) relies on the fact that since ECPR-based fees perpetuate high prices, they imply a consumers' surplus loss and a social welfare loss in comparison to prices that would reflect social economic costs and not private opportunity costs. Economides and White (1995, 1998) show that the consumers' surplus loss associated with the application of the ECPR can easily outweigh any welfare losses from productively inefficient entry. This means that (1) if the entrant is productively efficient, application of the ECPR just results in consumers' surplus loss and social welfare loss; (2) if the entrant is productively inefficient, for many parameter values, application of the ECPR results in consumers' surplus loss and social welfare loss.

7. Why Are the ECPR and M-ECPR Inefficient?

The ECPR and the M-ECPR guarantee the incumbent the recovery of its full pre-entry profits, all the way up to the full monopoly level, as well as full recovery of any historic cost inefficiencies under which the incumbent is laboring.[6] That is, the ECPR fee $p_A = p_{AB} - c_B$ that the entrant pays for monopolized component A contains the incumbent's full pre-entry profits *and* any cost inefficiencies, since these are part of the pre-entry final price p_{AB} and are not part of cost c_B. Moreover, if the incumbent's pre-entry production was inefficient, such inefficiency would push the ECPR fee higher, since it would be reflected in p_{AB} and, through it, in p_A.

The M-ECPR is inappropriate for pricing unbundled network elements and services because it leads to allocative and dynamic inefficiencies when the final service price is above incremental cost, through the perpetuation of high prices for final services. Accordingly, consumers who would have been served in a competitive market in the absence of M-ECPR (for example if the monopolized bottleneck were offered at cost) are excluded from the market because of the high price that results when M-ECPR is applied. This leads to allocative inefficiency. This means that, although the forward-looking economic costs of the final service are below consumers' willingness to pay, the application of M-ECPR-based access fees results in artificially high costs for providers of the service that is complementary to the bottleneck, as well as artificially high prices for the final service. As a result, some consumers are excluded from purchasing the service, and others buy less of it, resulting in a social welfare loss.

8. Inefficiencies Resulting from the Application of the ECPR and the M-ECPR

Various inefficiencies result from pricing bottleneck facilities according to the ECPR or the M-ECPR. We go through these on a case by case basis.

8.1 The Case Where the Entrant(s) in the Complimentary Goods Market is Equally or More Efficient than the Incumbent

When there are more or equally efficient potential entrants in the complementary component market, application of the M-ECPR results in a pure allocative loss. The M-ECPR keeps the price of the bottleneck facility high, entry occurs, and final prices remain high to the detriment of consumers. Application of the ECPR or the M-ECPR nullifies all the potentially positive effects of entry on consumers and social welfare.

[6] There is no requirement that the pre-entry market be a monopoly for this statement to be true, as Sidak and Spulber (1996) incorrectly assert. The ECPR and M-ECPR guarantee to the incumbent *any* supernormal profits that it was making before entry, thus perpetuating an allocative inefficiency whenever the original (pre-entry) final service price is above incremental cost. And, clearly, the final service price can be below the monopoly level but above incremental cost

8.2 Using the ECPR and M-ECPR to Exclude Equally Efficient or More Efficient Rivals when there Are Increasing Returns to Scale

The ECPR and the M-ECPR have been proposed as a way to exclude entry in the complementary good market by *inefficient* entrants. When the technology of production exhibits increasing returns to scale, however, the monopolist can use the M-ECPR to exclude or marginalize an equally or more efficient rival. The bottleneck monopolist may use the M-ECPR to establish high interconnection or access charges that result in a restriction of the scale of operation of the rival in the complementary market. Given increasing returns to scale, the rival then operates at the high end of its average cost curve. Thus, the monopolist is able to raise the production costs of its rival through the application of the M-ECPR. This results in a competitive disadvantage for the rival, as well as in higher prices for final services. Consumers are deprived of lower prices that would have resulted from competition if M-ECPR were not applied, as well as of competitive choices.

8.3 The Case where the Entrant(s) in the Complimentary Goods Market is (are) Less Efficient than the Incumbent

An allocative inefficiency can also result from the application of the M-ECPR when rivals are less efficient than the incumbent. Imposition of an M-ECPR access fee would not allow entry by firms that are less efficient than the incumbent. However, if the M-ECPR were not applied and use of the bottleneck facility were offered at incremental cost, often the entry of even inefficient rivals will decrease the final service price to the benefit of consumers. Thus, if the M-ECPR were not applied, entry would increase allocative efficiency. This increase in allocative efficiency can yield higher overall economic efficiency, despite a decrease in productive efficiency. Thus, even if the potential entrant in a complementary good market is less efficient than the monopolist, application of the ECPR or M-ECPR often leads to overall efficiency losses, as shown for a wide range of parameters in Economides and White (1995, 1997).

For example, suppose that the incumbent's unit costs are \$0.50 for component A and \$0.40 for component B, and final service AB was originally sold at \$1.20, *i.e.*, at a profit of \$0.30 per unit. Under the M-ECPR, an entrant would have to pay \$0.80 (= 0.50 + 0.30) for component A. If an entrant is less efficient than the incumbent in service B, *i.e.*, if its production costs are above \$0.40, the entrant cannot enter and survive. Under the M-ECPR, there is no entry and the final service price remains at \$1.20. However, if component A were sold to the entrant at cost, i.e., at \$0.50, an inefficient entrant could enter, survive, and precipitate a significant final price reduction and an increase in allocative efficiency. For example, entry by an inefficient entrant with production costs for component B of \$0.41 per unit (above the efficient level of \$0.40) would result in a reduction of final price to \$0.91 (= .0.50 + 0.41). In summary, when component A is sold at cost (rather than the M-ECPR access fee that includes monopoly profit), the price falls by \$0.29, and this results in an allocative efficiency gain: more consumers buy the good, and the ones who used to buy, now buy more of it.

The effects of application of the ECPR are shown in Figures 1 and 2. Figure 1 shows the benefits of not applying ECPR when the entrant is equally efficient as the incumbent. Under

ECPR, the price remains at \$1.20 despite entry, while with cost-based pricing, price falls to \$0.90.

Figure 2 shows the benefits of not applying ECPR when the entrant is less efficient than the incumbent. Under ECPR, the entrant is foreclosed and the price remains unchanged. Under cost-based pricing, the entrant enters, and the price falls to \$0.91.

Figure 1. Benefits of Not Applying ECPR When Entrant is as Efficient as Incumbent

Incumbent

c_A = \$0.50	c_B = \$0.40	profit = \$0.30

Output price: p_{AB} = \$1.20.

Efficient Entrant Under ECPR

ECPR-implied fee for A = \$0.50 + 0.30 = \$0.80	$c_{B'}$ = \$0.40

Output price: $p_{AB} = p_{AB'}$ = \$1.20.
No price decrease.

Efficient Entrant With Cost-Based Pricing And No Application Of The ECPR

$p_A = c_A$ = \$0.50	$c_{B'}$ = \$0.40	↑

Output price: $p_{AB'}$ = \$0.90.
Price decrease = \$0.30.

Figure 2. Benefits of Not Applying ECPR When Entrant is Less Efficient than Incumbent

Inefficient Entrant Under ECPR

ECPR-implied fee for A = \$0.50 + 0.30 = \$0.80	c_B = \$0.40

Output price: p_{AB} = \$1.20.

Entrant is foreclosed. Price remains unchanged.

Inefficient Entrant With Cost-Based Pricing And No Application Of The ECPR

$p_A = c_A$ = \$0.50	$c_{B'}$ = \$0.41	↑

Output price: $p_{AB'}$ = \$0.91.
Price decrease = \$0.29.

8.4 Inefficiencies Arising from the Application of the M-ECPR when the Monopolist's Costs Are Not Well-Known or Not Transparently Observable

The M-ECPR access fee to the bottleneck facility is equal to the price of the final service minus the incumbent's incremental cost of the complementary component. Thus, when the

M-ECPR is applied, the monopolist has an incentive to understate its incremental costs of the production of the complementary component (i.e., the service where it faces competition) and then employ the M-ECPR to levy an exclusionary access fee to its rival. This results in higher incremental costs even for a rival that is equally or more efficient than the incumbent. Thus, *even equally efficient or more efficient rivals can be excluded.* In this case the ECPR and the M-ECPR fail in their declared objective, which is to exclude only inefficient entrants.

In the example, the actual costs of the incumbent were $0.50 for component A and $0.40 for component B, and final service AB was originally sold at $1.20, with a profit of $0.30. The M-ECPR access fee for A is $0.80 (= 0.50 + 0.30 = 1.20 - 0.40). If costs are unobservable from the outside, the incumbent has an incentive to claim that the cost of its B component is lower (and consequently that its original profit rate is higher). For example, the incumbent can claim that the cost of B is $0.36, which is 10% lower than its actual cost of $0.40. This results in a profit of $0.34 (= 1.20 - 0.50 - 0.36), and an M-ECPR access fee of $0.84 (= 0.50 + 0.34 = 1.20 - 0.36). Facing this access fee, entrants that are equally efficient in the production of B as the incumbent, i.e., with cost of $0.40, are foreclosed from entering since they can only sell AB at $1.24 (= 0.84 + 0.40), while the entrant still sells at $1.20.

Even if the monopolist were constrained not to earn positive economic profits in the bottleneck market, if its costs are not perfectly observed, it can claim that some incremental costs of the complementary services are incremental costs of the bottleneck service. Lower incremental costs of the complementary component justify a higher charge for the bottleneck under the M-ECPR. Again, this higher charge will deter even those rivals which are more efficient than the monopolist in the production of the complementary component.

8.5 Application of the M-ECPR Can Result in Inefficient Entry

Application of the M-ECPR may also result in inefficient entry. The M-ECPR keeps prices for the monopolized input artificially high. High prices invite entry in the monopolized bottleneck market (good A). Since entrants will have to beat the monopoly price (rather than the competitive price for that element which would equal its cost), even entry by firms that have higher production costs for A than the incumbent may be profitable and may occur.

8.6 Perverse Comparative Statics under the ECPR and the M-ECPR

In a competitive market, reductions in the costs of either A or B would be reflected in reductions in the prices of these components and in a reduction of the price of the final service. Under the M-ECPR, a reduction of the cost of the monopolized component A has no effect on the final price and on the access fee for A. Even more perversely, under the M-ECPR, a *reduction* in the incumbent's cost for component B results in an *increase* of the access fee that an entrant has to pay for A! In the example, when the cost of B falls (say because of technological change) from $0.40 to $0.36, this results, as shown earlier, in an access fee for A of $0.84 (= 1.20 - 0.36), which is higher than the access fee of $0.80 (= 1.20 - 0.40) that prevailed when the cost of B was higher.

8.7 Dynamic Inefficiency of the M-ECPR

One of the fundamental properties of a competitive environment is free and unencumbered entry. Free entry drives the competitive process and leads markets to lower prices as well as efficient production. On the other hand, the M-ECPR preserves the market power of an incumbent monopolist. It increases the costs of entrants in the good that is complementary to the bottleneck by acting as an "entry tax," i.e., an extra cost that entrants have to pay to enter. When the M-ECPR is applied, entrants face a smaller potential profit margin (difference between consumers' willingness to pay and cost). Thus, the M-ECPR diminishes entry and competition in the complementary goods market, and results in higher prices for final services.

Starting from a situation of inefficiently high prices for final services, application of the M-ECPR preserves these high prices by setting prices for intermediate goods at inefficiently high levels. Thus, competition is not allowed to take its course and drive prices to competitive, efficient levels. Application of the M-ECPR prevents the natural tendency of the market to self-regulate itself through competition.

8.8 Administrative Problems of the M-ECPR

It is prohibitively complex and difficult to apply the M-ECPR in practice. The M-ECPR defines an access fee for the monopolized service of the incumbent that includes the private opportunity cost to it resulting from entry in a complementary market. In telecommunications there are hundreds of goods/markets that are complementary to a typical monopolized bottleneck such as the "local loop." These markets correspond to various vertical services that LECs currently provide, as well as to various geographic locations of intraLATA, long distance, and international destinations where calls originating from that loop may terminate. It is clear that these markets have different prices and various price discrimination schedules are used. Adoption of the M-ECPR requires that a different access fee be determined for the same good or service, say access to the local loop, depending on its use in combination with other goods and services. This means hundreds of different M-ECPR-derived access fees for the same good -- clearly an unmanageable situation (as well as discriminatory). Further, even if a regulatory agency is able to untangle this chaotic situation and assign correctly the hundreds of M-ECPR-based prices to the same bottleneck service or unbundled network element, it would be almost certainly of no use. In the presence of different prices for the same service, there are very strong incentives for providers of complementary unbundled network elements and services to rearrange their own pricing structure so as to arbitrage the price of the monopolized element to a single price. And, that price will have, in general, little discernible relation to the hundreds of M-ECPR-based prices that the administrative agency would have worked so hard to set. In short, administratively, M-ECPR-based prices are unworkable and unlikely to last. It is much easier, as well as more efficient, to price unbundled network elements without reference to private opportunity costs.

9. Application of the M-ECPR and the ECPR in the Implementation of the 1996 Act

The intent of the Telecommunications Act of 1996 was to bring competition to *all* telecommunications markets. In particular, the 1996 Act attempts to bring competition to the monopolized local exchange markets. To facilitate competition as well as to neutralize the ILECs' current grip on local exchange markets, the 1996 Act allows for two new ways of entering the local exchange, besides the obvious way, in which entrants build their own facilities.[7] In the first new way of entry, an entrant can enter the local exchange market by leasing unbundled network elements (UNEs) from the ILEC at cost-based prices. In the second, an entrant can buy at wholesale prices any service that the ILEC provides and resell it. This method of entry allows competition only for the retailing of telecommunications services. On the other hand, entry through UNE leasing should allow the entrant to produce all network functions and telecommunication services.

Currently, given the ILECs' virtual monopoly in markets for UNEs, one cannot rely solely on the market to determine economically efficient prices for UNEs. The 1996 Act correctly attempts to imitate a market environment by mandating the setting of cost-based prices for UNEs. To be economically efficient, prices have to reflect the forward-looking economic (minimized) cost of a present-day, telecommunications-only, efficient network. Such a cost measure is the Total Element Long Run Incremental Cost ("TELRIC") of a network element.

We show below that pricing rules that are based on private opportunity costs, such as the ECPR and the M-ECPR are inconsistent with the 1996 Act.

9.1 The M-ECPR and the ECPR Are Inconsistent with the 1996 Telecommunications Act ("Act" or "the Act")

A key requirement of the Act is that prices for unbundled network elements be based on cost (Act at Section 252(d)(1)(A)(i)). The M-ECPR fails to satisfy this requirement and is therefore inconsistent with the Act because, in my view, the correct meaning of "cost" in the Act does not include "private opportunity cost." Another requirement of the Act is that prices be non-discriminatory (Act at Section 252(d)(1)(A)(ii)). The M-ECPR however, implies different prices for the same unbundled network element, depending on with which complementary component or service it will be combined to create a final service. Given the wide diversity of components and services that can be combined with the same network element, M-ECPR-based prices for the same unbundled network element are, in general, different. In other words, the M-ECPR implies discriminatory pricing for the same component (which is produced at the same cost), contrary to the Act. Only the setting of a single price is consistent with the Act, and is inconsistent with the logic of M-ECPR.

As in the previous example, let component A have a unit cost of \$0.50, component B a unit cost of \$0.40, and let final service AB be sold at \$1.20. As we have shown earlier, this implies an M-ECPR-based fee for A of \$0.80 (= 1.20 - 0.40) when used as part of AB. Now let a second final service be created by combining A with component C. Let the cost of

[7] Act at section 252(d)(1)(A)(i).

component C be $0.60 and let the final product AC be sold at $1.50. Then the M-ECPR-based fee for A when sold as part of AC is $0.90 (= 1.50 - 0.60). A firm that sets prices according to the ECPR or M-ECPR, therefore, will charge $0.80 for component A to anyone who buys it for use in combination with component B, but will charge a higher price, $0.90, to anyone who uses A in combination with component C. This is a clear violation of the Act's non-discrimination requirements. On the other hand, the adoption of a single price for good A would violate the logic of the ECPR and the M-ECPR.

10. FCC's Position on the Appropriateness of the ECPR as a Costing and Pricing Principle

The FCC in its First Report and Order on the implementation of the 1996 Act concluded (at paragraph 709):[8] "ECPR is an improper method for setting prices of interconnectionn and unbundled network elements because the existing retail prices that would be used to compute incremental opportunity costs under ECPR are not cost-based. Moreover, the ECPR does not provide any mechanism for moving prices towards competitive levels; it simply takes prices as given." In further criticism of ECPR, the FCC concluded: "The ECPR, however, will serve to discourage competition in these very markets because it relies on the prevailing retail price in setting the price which new entrants pay the incumbent for inputs. While ECPR establishes conditions for efficient entry given existing retail prices, as its advocates contend, the ECPR provides no mechanism that will force retail prices to their competitive levels." (Order at Paragraph 710). The FCC's arguments hold equally for the M-ECPR.

11. Multiple Cost Recovery under the M-ECPR

If an ILEC uses the M-ECPR to set prices for unbundled network elements at stand-alone cost, this would that lead to multiple recovery of shared and common costs. Stand-alone costs include shared and common costs. If unbundled network elements are sold at stand-alone costs, the ILEC is able to recover the shared and common costs many times -- once for each element.

Multiple recovery of shared and common costs is inconsistent with the FCC's First Report and Order.[9] The First Report and Order strictly forbids multiple recovery of shared and common costs. Thus, the M-ECPR would lead to a clear violation of the FCC's rules. Further, the Doane *et al.* (1996) claim that an ILEC should be able to recover its historical costs (page 3). In contrast, the First Report and Order clearly dismisses the notion of using book or embedded costs. Instead, the FCC advocates the use of forward-looking, efficient costs.

[8] See Federal Communications Commission (1996).

[9] See Federal Communications Commission (1996).

12. Basing UNE Prices on Historical or Embedded Costs Would Lead to Inefficiency

Historical or embedded costs of a network element or of a service do not reflect the current cost to produce the element or service. This is because historical costs do not reflect current prices of inputs used in the production of the element or service and because they are not calculated using the most efficient current production technology. Thus, it would be economically inefficient to base UNE or services prices on embedded or historical costs.

Moreover, many of the historical costs of the ILEC's network are sunk. Thus, using M-ECPR to set final services prices that include these sunk costs will further tip the playing field in favor of the incumbent and create an unfair advantage. Deriving wholesale service prices and/or unbundled network element prices through the application of M-ECPR to set final services prices that include historical sunk costs will foreclose from the complementary goods market providers that are equally or more efficient than the bottleneck monopolist. Therefore, two types of efficiency loss would result from using M-ECPR to set final service prices based on historical costs: first, an allocative efficiency loss, since prices are held above TSLRIC; and, second, a productive efficiency loss, since the most efficient producer is foreclosed from the complementary goods market.

13. The Issue of Stranded Revenue

The overwhelming dismissal by state regulatory commissions and by the FCC of ECPR and M-ECPR, as well as those authorities' basing of prices on forward looking, rather than historical costs, lead the proponents of M-ECPR to attempt to recover ILECs' embedded or historical costs through a claim that the Act creates "stranded assets."[10] The claim is that the 1996 Act introduced competition, which results in lower market share for the ILEC and lower market prices. Doane *et al.* (1997) claim that this makes the assets of ILEC "stranded," and ask for a "transition charge" imposed on consumers and on entrants to "make the ILEC whole."

Doane *et al.* (1997) provide no evidence of particular "stranded assets." Even if some services were achieving a profit rate below the competitive rate of return, since typically each asset is used in the production of many services, it is unlikely that the *asset* would be realizing a profit rate below the competitive level of return. The nature of the proposal is to find ways to collect the ILEC's pre-Act revenue. Thus, it is not really about "stranded assets," but about "stranded revenue."

The attempt to collect a "transition charge" to recover "stranded revenue" is a clear attempt to create artificial barriers to entry and to harm competition. Competitors would have to pay the ILEC a "ransom" for every customer they gain. Keeping the ILEC's revenues and prices at pre-Act levels robs consumers of the benefits of competition. Moreover, the ILEC is encouraged to be inefficient, since it will have the same revenue irrespective of the level of its sales or its efficiency. Clearly, the proposal creates special advantages for GTE and is unfair to competitors. The proposal to collect "transition payments" from CLECs for reductions of

[10] See Doane *et al.* (1997), submitted in Hawaii as an attachment to the testimony of David Sibley on behalf of GTE.

the ILEC's share is clearly anti-competitive and directly opposes the objective of the 1996 Act to introduce competition in the local exchange. The proposal to collect so-called "transition charges" from consumers, if implemented, would rob consumers of the benefits of competition and perpetuate existing inefficiencies. In summary, the Doane *et al.* (1997) proposal has significant anti-competitive consequences.

14. Concluding Remarks

We have shown that access fees based on the ECPR, the M-ECPR, or other forms of private opportunity cost, perpetuate inefficiencies. Such fees include monopoly profits and historical cost inefficiencies, if they were present to start with, thereby resulting in losses in consumers' surplus and social welfare. In particular, prices based on the ECPR, the M-ECPR, or other forms of private opportunity cost are inappropriate for pricing unbundled network elements in the implementation of the Telecommunications Act of 1996. Prices for unbundled network elements of the telecommunications network that are based on the ECPR and the M-ECPR would create artificial barriers to entry and impede the emergence of competition. This is contrary to the intent of the 1996 Telecommunications Act.

References

Baumol, William J., (1983), "Some Subtle Pricing Issues in Railroad Regulation," *International Journal of Transport Economics,* vol. 10, no. 1-2, (August 1983), pp. 341-355.

Doane, Michael J., David S. Sibley, J. Gregory Sidak, Daniel F. Spulber, and Michael A. Williams, (1996), "An Economic Framework for Implementing the Pricing Provisions of the Telecommunications Act of 1996," unpublished manuscript.

Doane, Michael J., Paul W. MacAvoy, David S. Sibley, and Michael A. Williams, (1997), "An Economic Framework for Estimating Stranded Costs and Implementing a System of Recovery Through Competitive Transition Charges," unpublished manuscript.

Economides, Nicholas, (1996), "The Economics of Networks," *International Journal of Industrial Organization*, vol. 14, no. 2, pp. 675-699.

Economides, Nicholas, Giuseppe Lopomo and Glenn Woroch, (1996), "Regulatory Pricing Policies to Neutralize Network Dominance," *Industrial and Corporate Change*, vol. 5, no. 4, pp. 1013-1028.

Economides, Nicholas and Lawrence J. White, (1995), "Access and Interconnection Pricing: How Efficient is the Efficient Components Pricing Rule?," *The Antitrust Bulletin*, vol. XL, no. 3, (Fall 1995), pp. 557-579.

Economides, Nicholas and Lawrence J. White, (1998), "The Inefficiency of the ECPR Yet Again: a Reply to Larson," forthcoming, *The Antitrust Bulletin*.

Federal Communications Commission, (1996), "First Report and Order, CC Docket N. 96-98, CC Docket No. 95-185," August 8, 1996.

Laffont, Jean-Jacques and Jean Tirole, (1994), "Access Pricing and Competition," *European Economic Review*, vol. 38, pp. 1673-1710.

Sidak, J. Gregory, and Daniel F. Spulber (1996), "The Tragedy of the Telecommunications: Government Pricing of Unbundled Network Elements Under the Telecommunications Act of 1996," *Columbia Law Review*, vol. 97, no. 4, pp. 1081-1161.

Willig, Robert D., (1979), "The Theory of Network Access Pricing," in Harry M. Trebing, (ed.) *Issues in Public Regulation*. Michigan State University Public Utilities Papers. Proceedings of the Institute of Public Utilities Tenth Annual Conference.

Chapter 8

THE $500 BILLION OPPORTUNITY: THE POTENTIAL ECONOMIC BENEFIT OF WIDESPREAD DIFFUSION OF BROADBAND INTERNET ACCESS*

Robert W. Crandall
Brookings Institution
Charles L. Jackson
Jackson Telecom Consulting

I. INTRODUCTION

The stunning surge in economic growth in the late 1990s may not be sustainable if the U.S. economy does not encourage the deployment of the infrastructure required to continue the information technology (IT) revolution. As we went from personal computers to *networked* personal computers, our standard phone connection to the Internet was initially sufficient. But today we need much faster connections to allow us to exploit the ever-expanding opportunities of Internet connectivity. This chapter provides a rough estimate of the potential that could be unleashed by more rapid connections – by the evolution to a *broadband* environment.

We conclude that the universal adoption of broadband Internet connections by U.S. households could eventually provide consumers with benefits in the range of $170 billion to $280 billion per year. Moreover, producers of networking equipment, household computers, ancillary equipment, and software, and producers and distributors of entertainment products could also benefit by more than $100 billion per year.

---

* The authors are grateful to Verizon Communications for its support for an earlier version of this study.

II. Information Technology and U.S. Economic Growth

The vibrant performance of the U.S. in the 1990s has spurred a great deal of research on the sources of economic growth, especially the sources of differential growth between the first and second half of the 1990s. In the first half of the decade, GDP grew at an annual rate of 2.2 percent, whereas in the second half the average annual growth rate was 3.7 percent per year.[1] At the same time, the estimated annual rate of price decline for computers more than doubled in the second half of the decade, from 15.8 to 32.1 percent.[2] A consensus is now developing among economists that the surge in economic growth is attributable to investment in information technology, which in turn is attributable to the price decline in information technology equipment.[3]

A. Technical Progress in the Information Technology Sector

The revolution in electronics that has engulfed us is obvious by now. The prices of computers, monitors, printers, fax machines, and telephone terminal equipment have been falling dramatically. The prices of telecommunications networking equipment – such as switching gear and transmission equipment – have fallen at similarly spectacular rates. These price declines for information technology equipment can affect GDP growth in two ways. The first is through capital deepening. Lower prices for information technology equipment lead producers to substitute capital for labor, which results in more capital per unit of labor and makes labor more productive. The second is through productivity growth. The new information technology allows producers to use networking to employ more efficient means of production, thereby leading to higher levels of output per each unit of input. This second process is referred to as *technological change* and is generally measured in terms of productivity growth. Professor Erik Brynjolfsson of M.I.T. and Professor Lorin Hitt of the University of Pennsylvania conclude that technological change is likely to have been more important than capital deepening during the 1990s.[4]

The dramatic declines in prices for information technology equipment are obviously important for the economy because they provide businesses and consumers with more computing power at lower prices. Equally important, however, is the fact that these less expensive, more powerful devices allow for new applications of IT technology that were not even imaginable a decade ago. Today's broad Internet did not exist a decade ago. Consumers did not buy securities, books, or airline tickets by sitting at their home computers in 1990. Indeed, less than one in five households even had a computer in 1990.[5]

---

[1] U.S. Department of Commerce, Bureau of Economic Analysis, available at *www.bea.gov/bea/dn/gdplev.xls*.

[2] Dale W. Jorgenson, "Information Technology and the U.S. Economy," *American Economic Review*, 91(1), March 2001, Table 2.

[3] In particular, see Dale W. Jorgenson and Kevin J. Stiroh, "Raising the Speed Limit: U.S. Economic Growth in the Information Age," *Brookings Papers on Economic Activity*, 2000–1, (125-235) and the discussion following the paper.

[4] Erik Brynjolfsson and Lorin M. Hitt, "Beyond Computation: Information Technology, Organizational Transformation and Business Performance," *Journal of Economic Perspectives,* 14(4), Fall 2000, at 45.

[5] See Figure 2 below.

B. Measuring IT's Contribution to Economic Growth

Until recently, economists were puzzled by the absence of evidence that the revolution in electronics was contributing very much to economic growth. Now, however, such evidence is emerging, and it appears that IT *networking* is a crucial link. Professor Dale Jorgenson of Harvard University provides the most detailed decomposition of the sources of growth in GDP and total factor productivity.[6] Jorgenson concludes that the surge in growth in the second half of the 1990s is a result of the rapid decline in semiconductor prices, which led to falling IT equipment prices. The rate of price decline for IT equipment doubled in the second half of the 1990s, which caused massive investment in IT and, in turn, led to capital deepening and productivity growth.

Jorgenson disaggregates average labor productivity (output per hour) growth into growth in labor quality, capital deepening, and growth in total factor productivity. He shows that average labor productivity grew by 1.2 percent in the first half of the decade and by 2.1 percent in the second half of the decade. The growth in total factor productivity – the growth in output per unit of capital, materials, and labor – tripled between the first and second halves of the decade from 0.24 to 0.75 percent per year. The growth in total factor productivity due to IT doubled from 0.25 to 0.50 percent per year. Between 1995 and 1999, two-thirds of the growth in labor productivity was attributable to information technology.

In a separate investigation, Stephen Oliner and Daniel Sichel of the Federal Reserve Board conclude that information technology accounted for about two-thirds of the increase in labor productivity growth between the first and second halves of the 1990s.[7] Oliner and Sichel calculate that the information technology sector, which includes computers and semiconductors, accounted for two-fifths of the growth in total factor productivity between the first and second halves of the 1990s. They point out that "[t]hese are remarkable percentages given the tiny share of this integrated computer sector in total current-dollar output and they attest to the extraordinary pace of innovation in this part of the economy."[8]

A third study by two members of President Clinton's Council of Economic Advisers, Martin Baily and Robert Lawrence, confirms the findings of the Jorgenson and Oliner and Sichel analyses.[9] Baily and Lawrence show that there has been a substantial acceleration in total factor productivity outside the IT production sector since 1995. The authors conclude that even though one cannot definitively attribute the acceleration in total factor productivity growth to investment in IT, the fact that total factor productivity grew faster in IT-intensive industries suggests that the increase in productivity growth is mainly attributable to IT investment.

---

6 See Jorgenson, *supra* note 1.

7 Stephen D. Oliner & Daniel E. Sichel, "The Resurgence of Growth in the Late 1990s: Is Information Technology the Story," *Journal of Economic Perspectives*, 14(4), Fall 2000, at 21. An article by Gordon in the same issue argues that there is no productivity growth outside the IT production sector and that the investment in computer equipment has been unproductive. Oliner and Sichel show that the key difference between Gordon's and their analysis is the removal of what Gordon calls the cyclical factors in the growth of labor productivity. Oliner and Sichel conclude, "Whatever opinion one has of the particulars of Gordon's cyclical adjustment, the fact remains that his numbers embed our basic finding – that the production and use of information technology have contributed importantly to the actual pickup in productivity growth since 1995." (19).

8 *Id*, at 17.

9 Martin N. Baily & Robert Z. Lawrence, "Do We Have a New E-conomy," *NBER Working Paper 8243*, April 2001 (downloaded from http://www.nber.org/papers/w8243).

In another article, Professor Jorgenson and his co-author Kevin Stiroh of the Federal Reserve Bank of New York provide evidence that the surge in productivity is due to the effects of information technology outside the computer and software industries.[10] This contribution results from the use of the new technology to deliver goods and services, including the use of networking, in general and the Internet, in particular. Given the difficulty of measuring the prices of computers and semiconductors, Jorgenson and Stiroh provide three separate estimates of the contribution of IT technology to productivity growth for different assumed rates of price decline in IT. Their results are shown in Figure 1.

Figure 1. Contribution of IT and Non-IT Producing Industries to Total Factor Productivity

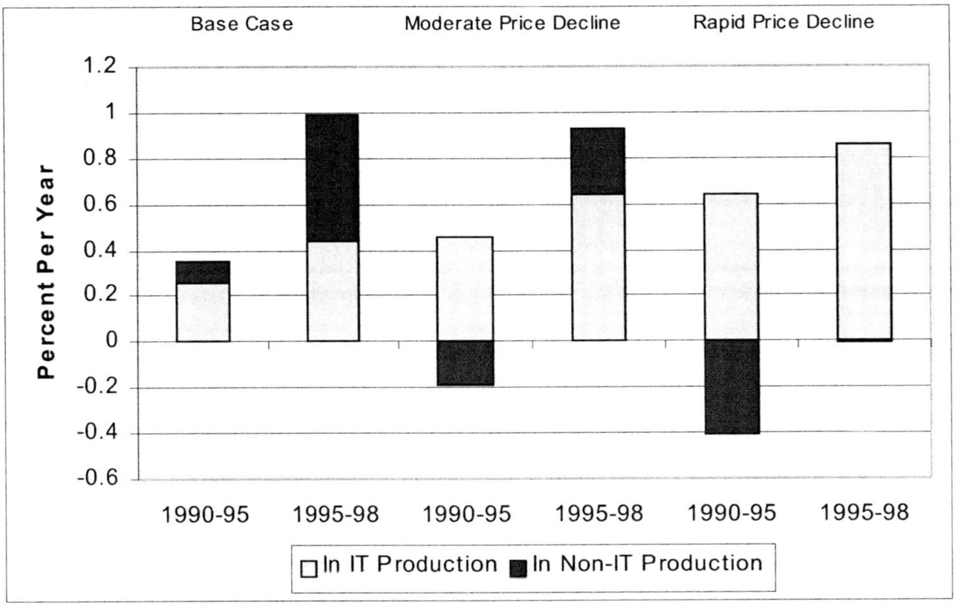

Source: Jorgenson and Stiroh (2000), Table 5.

Note that IT accounts for more than the *entire* increase in total factor productivity growth under the "rapid price decline" assumption. Given these results, Jorgenson and Stiroh suggest that these increases probably reflect the spillovers from IT to other industries. Always cautious, they offer the following important observation:

> If these productivity gains do indeed reflect spillovers from IT into non-IT industries, this would provide some evidence for the "new economy" argument.[11]

This "new economy" argument is at the center of the need for broadband deployment.

[10] Jorgenson and Stiroh, *supra* note 3.
[11] *Id.*, at 160.

III. The Contribution of Networks

The digital revolution began long before productivity began to climb in 1995. IBM began to sell digital computers in the 1950s. Apple introduced the first personal computer in 1977. Bill Gates and Paul Allen founded Microsoft in 1975. By the mid 1980s, personal computers were proliferating rapidly, and digital stored-program control telephone switches were replacing the older analog switches. Fiber optics began to spread in the telecommunications network in the 1980s.

Despite all of these remarkable advances in information technology, productivity growth continued to languish. Real GDP growth remained modest throughout the late 1980s and early 1990s. But then suddenly, the economy began to surge, fed by accelerating productivity. Something had clearly changed.

A. The Rise of the Internet

In the 1980s, the computer revolution was limited by the absence of *networking* capabilities. A personal computer can read and transform information faster than earlier models, but until it is connected to a source of information, an increase in processing power provides limited benefits. However, once one computer can access remote sources of information and transmit the information generated to other computers, its benefit expands enormously. This networking capability developed slowly within large businesses in the 1980s and – to a lesser extent – among small businesses. Until 1983, the predecessor to the Internet – the Arpanet – was limited largely to government and research institution use. Indeed, the TCP/IP protocol that is at the core of the current Internet was not even adopted by the Defense Advanced Research Projects Agency (DARPA) until 1983.

However, it was not until the 1990s that even large businesses routinely networked with other entities to exchange information or conduct transactions. The World Wide Web was born at the end of 1990 with the deployment of a server in a laboratory in Geneva, Switzerland.[12] As recently as 1990, few businesses even had e-mail.

B. Household Computer Use

For small businesses and individual households, the networking of computers for rudimentary access to information and e-mail communications did not develop in earnest until Internet use rose in the mid 1990s. The term *Internet* did not even enter the popular lexicon until 1994,[13] and few households or small businesses would even have known about the possibilities of such networking until at least 1994.

Home computer ownership accelerated in the mid 1990s as the Internet developed. Twelve years after Steve Jobs introduced the Apple II, only 12 percent of households had a

---

[12] *See* Tim Berners-Lee, "The World Wide Web: Past, Present and Future," (Aug. 1996), (downloaded from: http://www.w3.org/People/Berners-Lee/1996/ppf.html).

[13] *See* National Research Council, "Realizing The Information Future: The Internet and Beyond," (1994), (downloaded from: www.nap.edu/readingroom/books/rtif.html). This report provided the first popular dissemination of the notion of an "Internet."

computer. In the next twelve years, this percentage would more than quadruple to 56 percent (see Figure 2).[14]

Figure 2.

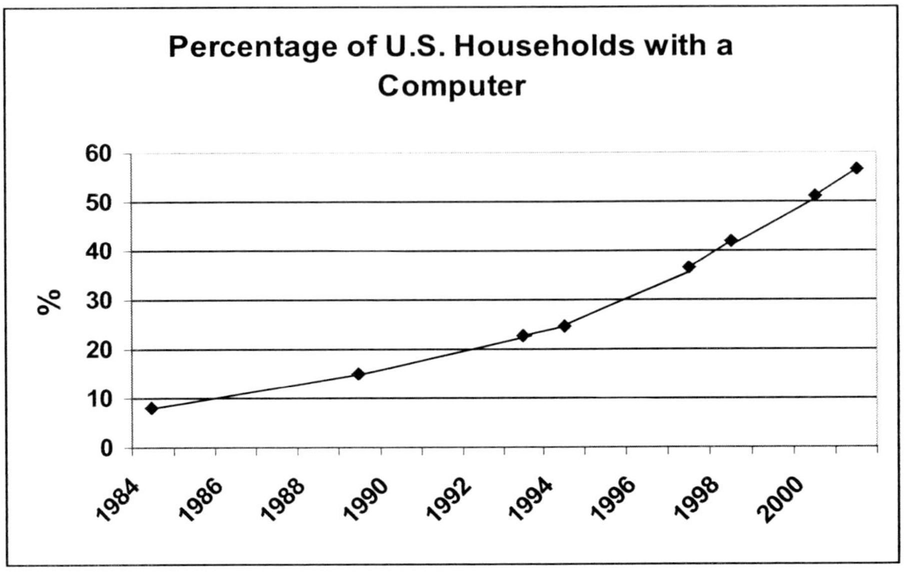

Source: U.S. Bureau of the Census, *Current Population Survey.*

C. The Growth of the Internet

At first, most households used the Internet for the most rudimentary of searches or for e-mail. Then e-commerce developed, allowing consumers to buy airline tickets, books, home electronic equipment, compact discs, and even securities on the Internet. But even these applications were limited by the slow speeds at which consumers accessed the Internet – dial-up speeds range between 2.4 and 56 kilobits per second. This may be sufficient to send simple e-mails and to download small documents, but hardly fast enough to download large audio or video files, to obtain complicated visual images required for real-time computer games, or to conduct advanced e-commerce. As these and other applications began to appear, households began to demand higher-speed Internet connections, but the network infrastructure for delivering them simply did not exist. Without this infrastructure, the evolution of the Internet to its full potential is temporarily blocked and its contribution to the acceleration in economic growth may soon come to an end.

The sharp rise and subsequent deceleration in the growth of Internet *use* is shown in Figure 3. Although household computer ownership has continued to grow and Internet penetration had risen to more than 50 percent of all households by 2001,[15] the growth in the average number of household hours on the Internet is slowing. It is possible that each new household uses the Internet less than early adopters, and the latter group's use is not

[14] Data obtained by the National Telecommunications and Information Administration from the Current Population Survey. See *www.ntia.doc.gov/ntiahome/dn/hhs/HHSChartsIndex.html.*

[15] See NTIA, fn. 15, *supra*.

continuing to grow sufficiently to maintain a constant growth rate. This is surprising given the new uses of the Internet that are appearing daily.

Figure 3. Average Hours of Internet Use

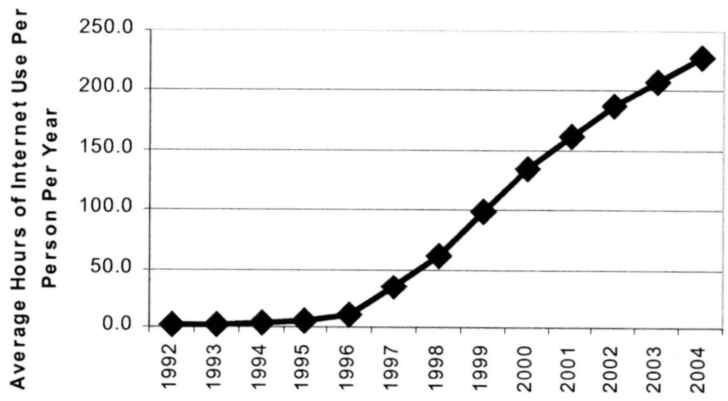

Sources: Bureau of the Census, Statistical Abstract of the United States, and Veronis, Suhler, and Associates, Telecommunications Industry Report.
Note: Figures for 2000 to 2004 are estimates.

The slowdown in the growth rate of Internet *use* is likely, in part, the result of users' frustration at trying to use innovative new network applications at the slow speeds allowed by ordinary dial-up connections. Unfortunately, only about 20 percent of households with Internet service subscribe to one of the new broadband technologies – DSL, cable modem service, or satellite service.[16] With so few broadband households, the new network applications we describe below will not develop fully.

IV. THE NETWORK REVOLUTION HAS JUST BEGUN

It is obvious that the Internet revolution has just started – after all, the Internet only appeared on the scene in the 1990s. Many in the computing industry now see a future of *pervasive computing.*[17]

The basic idea behind pervasive computing is relatively simple. As computing elements continue to proliferate in the environment, consumers will benefit if those computing elements can communicate with one another and with the outside world. The digital clocks around the house can set themselves to the right time by checking with a master clock. The furnace can measure its performance and determine that it needs to have its filters changed – and then e-mail the homeowner a reminder. Appliances can respond to time-of-day power

---

[16] Authors' estimate based on NTIA data. See fn. 15, *supra*.

[17] The National Institute of Standards and Technology (NIST) held a conference at the beginning of May 2001 on pervasive computing. The conference agenda was downloaded from: http://www.nist.gov/pc2001/agenda.html. A search on the web using the Google search engine yields 16,800 hits for the term "pervasive computing" and 16,600 hits for "ubiquitous computing."

prices or, in a world in which time-of-day metering is not implemented, to load-shedding commands from the authorities. Radios (radio-receiver-like appliances) can be tuned to any of thousands of Internet radio stations. If the home has a wireless local network, one can even carry such radios about in the home or listen to them over earphones while mowing the lawn – just as is done with traditional radios.

A. Opportunities with Network-Aware Products

The combination of broadband access and network-aware products will drive the market for home networking. It is quite possible that the most important benefits of pervasive computing cannot be foreseen. The same applies to key innovations in personal computing. No one forecast modern spreadsheets before Dan Bricklin developed the ideas that led to VisiCalc.[18] File sharing over the Internet was possible from almost the beginning using e-mail and FTP, but the development of the World Wide Web was a key factor in the growth of the Internet. No one forecast anything quite like the Web. Although the idea of linked hypertext had its predecessors in Engelbart's path-breaking NLS and Nelson's unimplemented Xanadu, the rapid adoption of the Web took almost everyone by surprise.[19] It appears highly likely that pervasive computing will lead to other innovations that are as revolutionary as spreadsheets or the Web.

Three elements of the communications infrastructure are needed to operate in most residences to support pervasive computing and the future innovations it will bring. The first is local networking – wire or radio communications connecting many devices in the home to the Internet. The second is an always-on connection from the home to the Internet. High-speed access is necessary for some applications, and it adds value to many others. The third requirement is a router or switch to connect the various devices in the home to the single connection to the larger Internet.

Networking inside the home presents issues that must be solved by individual consumers. Several technologies are possible – wireless, the existing telephone wiring, and the electrical power wiring. Products using these technologies are already on the market and are being used.

The need for a router can be met by installing the proper software in home computers or by building the function into the modem device. Microsoft Windows, the Linux operating system, and the Apple Mac all support the network routing functions needed to support many devices in the household from a single connection to the Internet.

The key issue of public policy concern is how to provide the incentives to deploy the new, higher-speed connection from the home or small office to the larger Internet. We discuss the nature of this connection below.

---

[18] *See* Dan Bricklin, "Was VisiCalc the First Spreadsheet?" (downloaded from: http://www.bricklin.com/ firstspread sheetquestion.htm).

[19] *See* Robert Cailliau, "A Little History of the World Wide Web", (1995), (downloaded from: *http://www.w3.org/ History.html*). One of the best forecasts of the social impacts of computer-communications systems was Parker, Edwin B. (1970) "The New Communication Media." *Toward Century: 21: Technology, Society, and Human Values*. New York: Basic Books.

B. Two Essential Attributes of Network Connections

The next generation of networking applications will rely heavily on two attributes of high-speed services – always-on and much higher speed. Each substantially increases the value of connections from the home or office to the Internet. For many applications, each of these attributes is essential – for other applications, they are only convenient.

Always-on refers to an Internet connection that is immediately available. ADSL, cable modems, some radio-based systems, and traditional high-speed T1 lines provide always-available connections. With an always-on or always-available connection, the delay from the time that a user goes to the computer and clicks on a web page icon to the time when the request for information is delivered to the remote server is measured in milliseconds. The alternative to always on is dial-up, or its equivalent. Using a dial-up connection to the Internet requires that a connection first be established, after which communications can proceed. Establishing a dial-up connection requires dialing a telephone call, connecting the telephone call to the terminating modem, synchronizing and training the modem, and logging in to the dial-in server. Typically, this process takes 30 to 45 seconds, or even more.

Without an always-on connection, some services, such as instant messaging, cannot be used until the user has taken the step of connecting to the Internet. Always-on connections are essential for certain applications, such as power load shedding and security services. Always-on connections facilitate other services, such as e-mail. With an always-on connection, e-mail software downloads messages before notifying the user that mail has arrived, allowing e-mail to be browsed at the speed of the local machine. Always-on connections also increase the use of the Internet. If one has a quick impulse to check on the Web for data, using a connection that is already established, rather than establishing a dial-up connection, improves the response time by about a factor of ten.[20]

High-speed refers to connections that are significantly faster than dial-up connections. Cable modems and ADSL modems provide access at data rates of 1 to 2 megabits per second – twenty to fifty times faster than typical dial-up connections.[21] The satellite services offered for home use today deliver data at about 500 kilobits per second – about ten times faster than a dial-up connection.[22]

C. Network Service Alternatives

Large organizations can obtain high-speed connections to the Internet using traditional telecommunications carrier offerings such as T1 lines, DS3 service, or Sonet services. The prices for these services are typically measured in the thousands of dollars per month – unaffordable for many small business locations and essentially for households.

Residential and small business users have several high-speed, always-on broadband service options, including cable modems, ADSL, and wireless services. These technologies are described in Appendix 1.

---

[20] The size of the improvement depends on how responsive the server is at the other end, the speed of the line, and the complexity of the page viewed.

[21] The authors are old enough to remember when the backbone of the Arpanet, predecessor of the Internet, ran at 56 kilobits per second and was considered high speed.

[22] Two firms, DirecPC and StarBand Communications, offer two-way satellite-based Internet access.

V. ECONOMIC BENEFITS FROM BROADBAND CONNECTIONS

Any new product or product improvement creates benefits for both consumers and producers. Consumers gain because they are able to purchase a new or improved product that was previously unavailable. They consume it up to the point at which the marginal value of the product to them is equal to its price. In the case of typical broadband services, consumers either subscribe to the service, or they do not. As the uses of broadband multiply, the value to subscribers rises far above the monthly subscription price. This is the *consumer surplus* from the innovation.

Producers of new services that rely on broadband, of products used in conjunction with broadband service, and even of the broadband service itself also gain from the greater diffusion of broadband. They bid resources away from other sectors of the economy and earn returns over and above those available elsewhere, until the marginal value of each type of resource is equalized across all alternatives. The *producer surplus* that is generated by inframarginal sales is a real benefit to producers and, therefore, to the economy.

In this section, we attempt to provide a rough estimate of the likely long-term gains to the economy – the sum of consumer and producer surplus – generated by widespread diffusion of broadband access. By *long-term,* we mean a time period sufficient for broadband to become virtually ubiquitous, given the appropriate policy environment – a time period that could stretch out to twenty-five years or even more.

A. Estimates of Consumer Value

We use two related approaches to estimating the potential benefits to consumers from the diffusion of broadband services. First, we estimate the direct benefits from the prospective demand for greater high-speed connectivity. Internet penetration grew rapidly between 1997 and 2000. (See Figure 4.) We assume that in 2003, 60 percent of households have residential Internet connections, and about 15 percent have broadband Internet connections. As broadband becomes *available* to more households, a larger share will use the Internet and thus create larger increases in consumer welfare than can be deduced directly from current estimates of the demand for broadband alone. Moreover, the demand for broadband will increase as new applications requiring high-speed connections are developed for Internet distribution.

Figure 4.

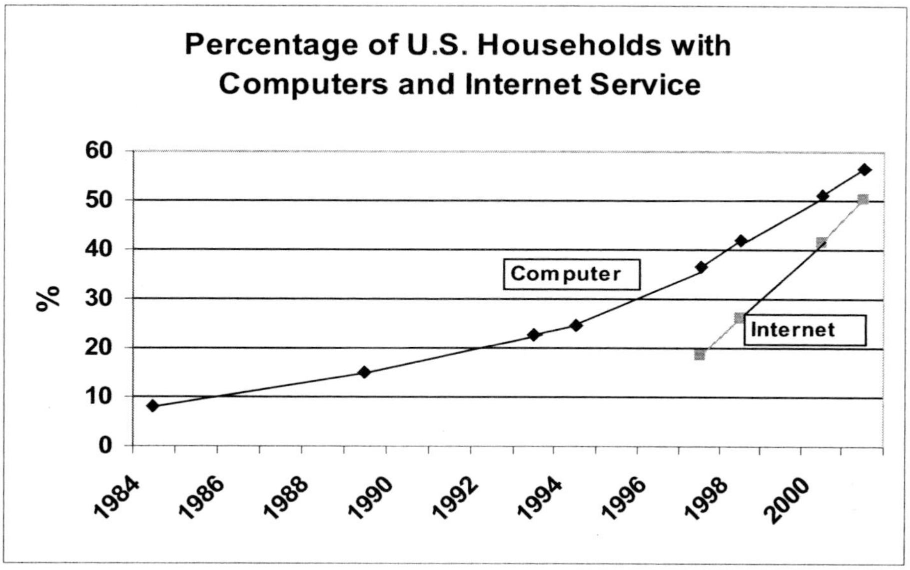

Source: National Telecommunications and Information Administration, downloaded from *www.ntia.doc.gov/ ntiahome/dn/hhs/HHSChartIndex.html.*

In addition, increasing the diffusion of broadband would lead to a greater household demand for personal computers and related devices because households would need faster computers with greater storage capacity to interconnect with services available at these higher speeds. The shift in demand for these products would create additional consumer surplus from non-broadband uses of this equipment that can be deduced directly from information on such demand.

Our second approach is based on indirect evidence of the potential value of the greater diffusion of broadband and more powerful home computing equipment by examining the new services that households could obtain from faster Internet connections. This approach involves estimating the increase in consumer welfare generated by the new services themselves, in addition to the savings in time and commuting that this new technology will allow.

B. Direct Estimate of Potential Consumer Surplus

The most straightforward estimate of the value of enhanced availability of broadband derives from information on consumer subscriptions to broadband services. We begin my assuming that about 15 percent of households subscribe to a broadband service; 60 percent subscribe to an Internet service of some kind; and 94 percent subscribe to ordinary telephone service.[23] Were broadband to become ubiquitous, it would resemble current telephone service in its household penetration.

[23] The number of broadband subscribers was 12.8 million as of December 31, 2001. Of these, 11.0 million were residences and small businesses. *See* FCC, *High Speed Services for Internet Access: Status as of December 31, 2000,* July 2002. We assume that residential broadband has grown to at least 15 million by the end of 2002.

1. An Estimate Based on Today's Demand Curve

Given current broadband penetration of 15 percent and an average price of the service of $40 per month, total broadband revenues may be estimated at $480 times 15.9 million, or $7.6 billion per year. Assuming that the demand for such service is linear with an elasticity of –1.0, the value of the service to these consumers – the consumer surplus – is $3.8 billion per year in addition to the $7.6 billion they pay. If the demand elasticity is –1.5, the consumer surplus falls to $2.5 billion.[24]

Were broadband to spread to 50 percent of households at $480 per month through a shift of a linear demand curve with constant slope, the annual expenditure on the service would rise to $31.2 billion.[25] At 50 percent penetration, the additional value to consumers would rise to between $42.5 billion and $63.8 billion per year at our two assumed two price elasticities.

If broadband service were to become truly ubiquitous, similar to ordinary telephone service, annual consumer expenditures on the service would rise to $58.7 billion per year, assuming the continued shift of the linear demand curve at constant slope and an annual price of $480. The additional value to consumers – over and above their expenditures on the service – would be $150.3 billion to $225.4 billion per year, assuming that the linear demand curve with a current elasticity of –1.0 or –1.5 simply shifted outward.[26]

The assumption of a shifting linear demand curve with constant slope is obviously arbitrary. A more modest estimate of consumer surplus would result from an assumption of log-linear demand with a "choke" price, *i.e.,* a price above which no one would subscribe, of, say, $120 per month.[27] As Table 1 shows, under this assumption, consumer surplus for 50 percent broadband penetration falls to between $26.5 billion and $34.4 billion per year, and for 94 percent penetration to between $49.8 billion and $64.4 billion. These more conservative estimates are based, however, on the assumption that the price elasticity of demand for a ubiquitous service, or "necessity," is equal to -1.0 to -1.5, surely far greater than the elasticity of demand for other necessities.

The estimates for Internet and telephone subscriptions are based on the Bureau of the Census, Current Population Survey.

[24] Some earlier estimates of the value of broadband were even be lower. For instance, using data on broadband connections in 1998–99, Austan Goolsbee of the University of Chicago finds that the consumer surplus from broadband services is only $700 million per year. *See* Austan Goolsbee, "The Value of Broadband and the Deadweight Loss of Taxing New Technology," *University of Chicago Working Paper*, (November 2000).

[25] These calculations assume it takes 25 years to reach universal broadband penetration. We assume a total of 106 million households at present and 130 million households in 2027. The calculations assume a 0.8 percent rate of growth, which is slightly more conservative than the Census projection of 1.08 percent.

[26] At this "ubiquitous" level of demand, the price elasticity of demand would be between –0.13 and –0.20, still substantially above the current estimates of the price elasticity of demand for telephone service but somewhat below current estimates of the elasticity of demand for dial-up Internet service. Note that linear demand curves with such demand elasticities imply that *someone* would be willing to pay as much as $353 to $510 per month for the service. This seems reasonable to us.

[27] This is the approach used by Roibert W. Crandall, Robert W. Hahn, and Timothy J. Tardiff, "The Benefits of Broadband and the Impact of Regulation," in Robert W. Crandall and James Alleman (eds.) *Should We Regulate Brosadsband?* Brookings, 2003.

Table 1. Estimated Increases in Consumer Surplus Under Different Assumptions about the Shape of the Demand Curve (Billions of Dollars)

Elasticity[a]	-1.0	-1.5
Linear Demand Curve		
Current Consumer Surplus	3.8	2.5
Forecasted Consumer Surplus at 50% Penetration	63.8	42.5
Forecasted Consumer Surplus at 94% Penetration	225.4	150.3
Demand Curve with Constant Elasticity and Choke Price of $120/mo.		
Current Consumer Surplus	8.4	6.5
Forecasted Consumer Surplus at 50% Penetration	34.3	26.5
Forecasted Consumer Surplus at 94% Penetration	64.4	49.8

Source: Authors' calculations

[a] In the case of the linear demand curve, the elasticity indicated refers to the elasticity at the current price and quantity demanded before the curve is shifted outward.

If broadband were to become as much of a necessity as ordinary telephone service, the demand for it would no longer be price elastic. Household demand for ordinary telephone service is extremely price inelastic. Recent estimates of this elasticity are -0.03 or even less in absolute value.[28] If broadband becomes as essential as ordinary telephone service is today, we would expect that the demand for it would become similarly price inelastic. As broadband becomes a "necessity" and, therefore, less price elastic in demand, its value rises sharply.

It is also likely that the real price of broadband service will fall over time, given the declining cost of electronics equipment. Thus, the value to consumers of the enhanced availability of broadband could easily be more than $225 billion per year, assuming that an outward shift of a linear demand curve from today's equilibrium is appropriate.

2. The Attractiveness of New Technologies Is Often Underestimated

The above methodology is obviously quite speculative. It is possible that current estimates of the demand for broadband underestimate the future demand for broadband because they are based on inferences drawn from the current demand for the service. Surely, consumers today do not have the range of uses for broadband that will be available in the future. Without such alternatives before them, consumers cannot provide us with evidence on how much they will value broadband. Indeed, in the past thirty years, market participants and analysts have frequently underestimated the prospective demand for new services. Prominent examples include the photocopier, cable television, and the wireless telephone.[29]

[28] For a survey of these estimates, *See* Lester D. Taylor, "Telecommunications Demand in Theory and Practice", Kluwer, (1994). For more recent estimates, *See* Robert W. Crandall and Leonard Waverman, *Who Pays for Universal Service: When Telephone Subsidies Become Transparent*, Brookings Institution Press, 2000, at Chapter 5.

[29] See Appendix 2 for details.

C. Additions to Household Computing Capacity

The expansion of household demand for broadband will create additional demand for computers and networked home appliances. Approximately 40 percent of all U.S. households do not have a computer.[30] Clearly, these households are not equipped to connect to the Internet at any speed. Of the 60 percent of households with computers, many will need to upgrade their equipment to obtain greater processing speed, more random-access memory, or greater hard-drive capacity. Still others will choose to buy more advanced equipment.

Eventually, it is likely that households will invest in multiple computers – a reasonable limit is about one computer per person in the household. Household networking equipment will be needed. Computers will require bigger disk drives. Today's 30 or 40 gigabyte drives will become commonplace. Noncomputer devices will be upgraded with connections to the Internet (as discussed above). Obviously, it is much easier for a consumer to add an MP3 player or Internet radio than to add an Internet-ready furnace to her household.

Therefore, the broadband-induced demand for household equipment must include additional or better computers and home networking equipment. About 60 to 80 million households have cordless phones today; computer networking should eventually enjoy a similar popularity.[31] There are, however, some difficult standardization problems that must be solved before we move to the widespread deployment of household devices with Internet connectivity.

Connections to the Internet place new demands on computers. It is not unusual to download hundreds of megabytes of audio and video clips. An e-mail message with some attached photographs of the grandchildren can take up five megabytes of storage. Downloaded video clips can be massive – up to a hundred megabytes for each minute of DVD-quality video. Households with older computers will find that their hard disks are quickly exhausted. Other households will find that the memory or processor of their computer systems no longer provides adequate service. Many such consumers will upgrade – either adding the resources they lack or, for the sake of simplicity, buying new computers.

Some households will purchase multiple computers, and many households will find that adding networking capabilities inside the household creates significant benefits. As discussed above, always-on networking creates the opportunity for household equipment to perform more subtle diagnostics, to interact with remote systems, and to present a more convenient interface to consumers.[32]

---

[30] See NTIA, fn. 15, *supra*, for 2001 data..

[31] Approximately 130 million cordless phones have been sold in the United States since 1995. Electronic Industries Association, "Factbook" (2000).

[32] For example, one can buy thermostats today that have many features – the settings vary by time-of-day and day-of-the-week. Settings can be overridden. Operating such thermostats is difficult. The display shows only a few characters and the controller requires navigating a complex tree of options. It would be far easier if the thermostat presented a web interface and allowed the user to set the time, date, and desired settings in a more natural and convenient manner. But, such a communicating thermostat can provide many other valuable features at hardly any additional cost. It can be programmed to check energy prices and to base its actions on the price and forecasted prices of energy (cool the house down fast before the price of electricity goes up or let the temperature rise until the price goes down), it can signal abnormal conditions, say household temperatures lower than 40 degrees or higher than 95 degrees to an outside monitor – thus avoiding circumstances that might lead to frozen pipes or health problems. It is reasonable to expect that, over the next two or three decades, always-on networking and home networks will lead to the installation of many devices in the home that take advantage of the network in the home and the connection to the Internet.

Personal consumption expenditures on computers, peripheral equipment, and software rose from $6.2 billion in 1987 to $34.5 billion (in current dollars) in 2000 but the rate of increase in nominal spending has declined substantially in the last few years.[33] Were broadband to diffuse widely through the population, the share of households with computers would rise and the number of computers per household would also increase.

A conservative estimate of broadband's stimulus on household purchases of IT equipment would be that U.S. household spending on computer equipment, peripherals and software would resume its 1991–95 rate of growth of 14.2 percent per year, rather than continuing at its 1995–2000 growth rate of 9.9 percent per year. If growth returns to its 1991-95 pace, total spending would be $190 billion by 2012, rather than $113 billion, an increase of $77 billion. Were the broadband revolution to accelerate household equipment expenditures by another 3 percent per year to 17.3 percent annual growth, the additional spending in 2012 would be $159 billion.[34]

The increase in consumer welfare from this expansion of demand due to new broadband services depends on the elasticity of demand for household computing and networking equipment and software. For instance, if the price elasticity is –1.0 ten years from now at the prevailing level of demand growth without ubiquitous broadband, and if the demand curve is linear, the $77 billion increase in expenditures would imply an increase in consumer surplus of $103 billion per year.[35] The $159 billion increase in expenditures would convey $271 billion per year in additional consumer surplus. Jerome Foncel of the University of Lille and Marc Ivaldi of the University of Toulouse estimate that the price elasticity of household demand for computers is –1.66.[36] At this elasticity, the additional consumer surplus is $62 billion and $163 billion, respectively.[37] Not the entire surplus is *in addition* to the previous estimates of the benefits from broadband. However, we might assume that one-third of this surplus from the new household networking equipment accrues from non-Internet services. Thus, our estimate of the *additional value* from new household equipment, over and above that conveyed through broadband connections, is between $20 billion and $54 billion once broadband becomes ubiquitous.

We could perform a similar analysis for other household equipment, such as the thermostat with a web interface. However, we expect that such equipment will be a weaker complement to high-speed networking than household computing, and consumer surplus in the former instance will not be as driven by broadband access, as will computers and networking equipment. Given this factor and the more speculative nature of these benefits, we have not tried to quantify them.

To reiterate, broadband access will stimulate the sales of both computers and network-enabled household devices. The growth of networked household devices, other than

---

[33] Bureau of Economic Analysis. "National Accounts Data", (downloaded from: http://www.bea.doc.gov/ bea/dn1.htm).

[34] *Id.* These calculations are in nominal dollars. Obviously, the rate of growth of the purchase of *real* computing power has been much greater.. In order to be conservative, we do not extend these projections past ten years.

[35] Technically, *consumption* of these services is related to the consumers' stock of equipment, not new additions. The additional purchases generate such surplus over several years following purchase. We do not attempt to project the growth of consumer expenditures on computers and related equipment past ten years.

[36] Jerome Foncel & Marc Ivaldi, "Operating Sustem Prices in the Home PC Market," *University of Toulouse Working Paper*, (May 2001), (downloaded from http://www.idei.asso.fr/English/ECv/CvChercheurs/ EcvIvaldi.html).

[37] The corresponding Hicksian compensating differential numbers are $60 and $140 billion.

computers and entertainment equipment, will probably be a slow process, driven in part by the natural pace of updating the expensive systems in the home. Ultimately, it will induce substantial expenditures for each household.

D. New Services Provided by Broadband Connectivity

We now turn to the more indirect estimates of the value of the new services that broadband will make possible in order to buttress the direct estimates above. Forecasting the future is always difficult. Undoubtedly, many of the benefits of broadband deployment will be applications that can only be seen dimly at present, if at all.

The most readily forecast economic benefits of broadband fall into four general areas – retailing, transportation, home entertainment, and health care – and we consider each below. This is not to say that there could not be equivalent sources of benefit from other services, such as education, but we simply lack the tools or vision to analyze them at this time.

1. Retailing/Wholesaling

Broadband services will provide enormous benefits to consumers by improving the delivery of goods and services. We have already seen the tip of this iceberg with Amazon.com and other Internet bookstores, which offer convenience, but not necessarily lower prices. Buying a book over the Internet saves the consumer the time it takes to go to the bookstore, find the book, wait in the checkout line, check out, and return from the bookstore. For most of us, the costs in time, effort, and transportation are significant. They can be reduced if one stops at the bookstore while shopping for other items – but the time and other resources generally required for book shopping are substantial.

Books are well suited for sale over the Internet. The key information needed to decide whether to buy a book, such as reviews, the table of contents, and perhaps a summary or excerpt, can be displayed efficiently even over a dial-up connection. Many types of brand-name manufactured goods, covered by the manufacturer's warranty, are just as suitable for sale over the Internet as are books. If one wants an HP printer, one will get the same equipment whether one buys it over the Internet or from a local retailer.

The Internet is rapidly replacing travel agents for consumer bookings of airline tickets, hotels, and automobile rentals. Banking and securities transactions are also shifting to the Internet. As the speed of Internet connections increases, consumers can survey their options for such services more quickly, view many of the alternatives, and effect transactions that would have taken hours on the telephone or in person. Broadband Internet connections thus provide a new option for retailing and product distribution. Retail e-commerce accounted for only about 1 percent of total retail sales, or about $26 billion, in 2000.[38] The transition to broadband retailing as an alternative will increase this share and, therefore, increase consumer choice and economic efficiency.

The value added by retail and wholesale trade accounts for about 16 percent of GDP. Wholesale trade in the United States contributed $674 billion in value added in 2000 (7

[38] U.S. Department of Commerce, "Estimated U.S. Retail E-Commerce Sales: 4th Quarter 1999 - 1st Quarter 2001", (downloaded from: *www.census.gov/mrts/www/mrts.html*).

percent of GDP), and retail trade contributed $894 billion (9 percent).[39] These retailing costs are in addition to the cost of consumers' time involved in shopping.

Consumers must travel to traditional retail establishments and compare the choices at one or more of these establishments. Additional time is required to select items, pay for them, and transport the goods home. Estimates suggest that the average adult spends about 30 minutes per day shopping for nongrocery items and about 15 minutes traveling for such shopping – a total of 45 minutes per day for nongrocery shopping.[40] The official accounting of retail trade's contribution to GDP does not include the time a consumer spends in these activities. It does, however, include the time a retail store's buyers spend going from supplier to supplier examining products and comparing prices. The disparate treatment of these two quite similar activities arises from the fact that one involves a market transaction, and the other does not.[41] Nevertheless, the time a person spends shopping is an important economic cost.

Assuming that the average person values his or her time at one-half the average wage of $14.50 per hour, the time spent in non-grocery shopping or traveling for non-grocery shopping is worth $5.44 per day[42] or $1,986 per year per person. Multiplying by total persons 18 years of age or older (approximately 200 million), we get a total of $397 billion per year worth of time spent on nongrocery shopping.[43]

Combining producer value added for wholesale and retailing together with our estimate of consumer time expenditures, retail activities account for annual costs of roughly $2.0 trillion per year. Even modest increases in the efficiency of an activity worth such a massive amount will result in substantial benefits. If speeding up the deployment of broadband improved the efficiency of this sector by only 3 percent, as compared with the savings under a slower scenario, the savings would total $60 billion per year. If such deployment improved efficiency by 10 percent, the savings would total $200 billion per year at current values. Assuming that the retailing/wholesaling costs are fully passed on to consumers, these savings accrue to consumers for *existing* consumption levels. If the demand elasticity for all consumer goods is –1.0, another $1 billion to $10 billion must be added for the consumer surplus gains from additional consumption. Thus, the total gains may be estimated at $61 billion to $210 billion under these cost-savings assumptions.

When universal broadband service is achieved, these gains will be much greater because of general economic growth. Assuming that the economy grows at an average rate of 2.5 percent per year for ten years, these savings would grow to $78 billion and $270 billion (current dollars), respectively.[44] These gains obviously depend on the share of households with broadband and the assumed savings available from improvements in the efficiency of

---

39 *See* Bureau of Economic Analysis, "Industry Accounts Data – Gross Domestic Product by Industry" (downloaded from www.bea.doc.gov/bea/dn2/gpoc.htm#1993-99).

40 *See* National Science Foundation, "Family Time Use Study: 1998-1999 Time Diaries", (downloaded from: http://www.webuse.umd.edu/data_analysis.htm).

41 Such disparities in the treatment of economic activities abound in our system of accounting for economic activities. If one pays a painter $1,000 to paint one's house, that thousand dollars shows up in the GDP. In contrast, if the same person buys $100 worth of paint and supplies and paints his or her house, the GDP measurements only show $100 of GNP. State and federal income tax are not applied to the $900 of labor and skill supplied by the homeowner.

42 Average hourly wage rate for September 2001, Bureau of Labor Statistics data, *www.bls.gov/cps/home.htm*.

43 If we were trying to model future effects with exactitude, we would use future population levels. However, our goal here is to provide a general calculation that provides a solid feel for the magnitude of benefits associated with more rapid broadband rollout. In this context, refining the population figures would be spurious precision.

44 In order to be conservative, we do not extend these projections past ten years.

retailing/wholesaling. Table 2 provides estimates of the consumer benefits under various alternative values of these two crucial parameters.

Table 2. Consumer Gains from Broadband's Contribution to e-Commerce ($ Billions per Year)

Share of Households with Broadband Access	Improvement in Wholesale/Retail/Shopping Efficiency				
	0%	5%	10%	15%	20%
10%	$0	$14	$27	$41	$54
25%	$0	$34	$68	$102	$136
50%	$0	$68	$135	$203	$270
85%	$0	$115	$230	$345	$460
100%	$0	$135	$270	$405	$540

As the share of households that have broadband increases, however, one should increase one's estimates of the potential savings because of the "network effects" inherent in developing new shopping sites.

2. Reductions in Commuting

Telecommunications can substitute for travel – an oft-repeated truism. Substantial telecommuting has been just around the corner for the last three decades. In fact, telecommuting has grown slowly but steadily; helped in part by technologies such as personal computers, modems, fax machines, and high-quality telephone service.

As we have gained experience with telecommuting, our vision of this activity has changed. Obviously, many jobs – such as a car repair technician or a receptionist – are not suited for telecommuting. But many other jobs are. With modern telephone systems, some customer service representatives can work from home. With broadband Internet access, essentially every customer service representative or call center operator could work from home. In fact, with modern PCs and broadband access, most knowledge workers can have the same access to information at home as they have at their desks. The modern view of telecommuting takes many forms. For some people, for example, someone with a significant physical disability, telecommuting may entail working almost entirely from home. For others, such as journalists, most days may be spent in the office, but a crash-writing project may be most effectively conducted from home.

a. The Texas Transportation Study

The primary direct benefits of telecommuting occur from the reduction in travel required by the employee and the reduction in infrastructure costs at the office. But a significant secondary benefit is a reduction in congestion costs. The Texas Transportation Institute's Urban Mobility Study reports estimates of the costs of traffic congestion in 68 urban areas.[45] Its 2001 report states:

[45] *See* Texas Transportation Institute, *supra* note 42.

> Congestion costs can be expressed in a lot of different factors, but they are all increasing. The total congestion "bill" for the 68 areas in 1999 came to $78 billion, which was the value of 4.5 billion hours of delay and 6.8 billion gallons of excess fuel consumed. To keep congestion from growing between 1998 and 1999 would have required 1,800 new lane-miles of freeway and 2,500 new lane-miles of streets – or – 6.1 million new trips taken by either carpool or transit, or perhaps satisfied by some electronic means – or – some combination of these actions. These events did not happen, and congestion increased.[46]

The TTI report shows that 80 percent of these $78 billion in costs occurs in only 24 cities (comprising most of the larger cities in the United States), and 90 percent occurs in 36 cities. In Los Angeles, traffic congestion imposes estimated costs of $1,000 per person per year. Of course, the marginal congestion – that created by one additional commuter – is much higher than the average. Examining these data, Professor Paul Krugman of Princeton University concludes, "Do the arithmetic and you find that each individual's decision to commute by car in Atlanta imposes congestion costs of $3,500 per year, or $14 per workday, on other people. These are costs over and above the costs actually paid by the driver himself – that is, they are costs that drivers don't take into account. And this number does not take into account environmental impacts (air quality in Atlanta is steadily deteriorating)."[47]

The benefits from modest increases in telecommuting are thus far greater than would be indicated by the average values of congestion costs. Krugman's analysis indicates that in the case of Atlanta, the marginal congestion cost is 3.8 times the average congestion cost.[48] Thus, a 1 percent reduction in busy-hour traffic in Atlanta could be expected to save almost $100 million per year.[49]

b. A National Estimate

The quantifiable benefits of telecommuting are the savings in transportation costs – both the time and expense of the worker and the reduction in congestion and pollution costs imposed on others. Non-quantifiable (or at least much harder to quantify) benefits of telecommuting include improvements in worker productivity and the expansion of employment opportunities for people with disabilities.

Various studies estimate that 20 to 40 percent of jobs permit telecommuting at least part of the time.[50] If we assume that 30 percent of jobs permit telecommuting an average of 20 percent of the time (1 day per week or 50 days per year) and that the average commuter trip

---

[46] *Id,* at iii.

[47] Paul Krugman "Nation in a Jam," *New York Times,* Op-Ed page, (May 13, 2001). Professor Krugman's statement must be read carefully. He is asserting that the average marginal congestion costs of a single added commuter are about $3,500 per year. It is well known that, in queueing systems, the marginal congestion externalities are far worse than the average externality.

[48] The average congestion cost was $915 per person. *See* Texas Transportation Institute, *supra* note 42, at Appendix 9, (downloaded from: http://mobility.tamu.edu/2001/study/tables/A9.pdf).

[49] TTI estimated total congestion costs in Atlanta of $2.62 billion. The calculation is (3.8) · (1%) · $2.62 billion = $99.6 million.

[50] *See* International Telework Association and Council, "Telework America" (2000), (downloaded from: *http://www.telecommuteorg/twa2000/research_results_key.shtml*) and U.S. Dept of Transportation, "Secretary Slater Challenges Federal Agencies to Increase Telecommuting Goals", (Oct. 24, 2000), (downloaded from: *http://www.dot.gov/affairs/2000/dot21200.htm*).

for a telecommuter is 20 minutes, then we can calculate the potential savings in travel costs and congestion. The savings in travel time are:[51]

(180 million civilian labor force) · (30% possible telecommuters) · (33 hours/year) · ($6.20/hour) = $11.1 billion per year.

Similarly, the savings in travel costs are:

(180 million civilian labor force) · (30% possible telecommuters) · (450 miles/year) · ($0.3/mile) = $7.3 billion per year.[52]

We use Krugman's model for congestion externalities described above, but we employ a lower multiple (2.0 rather than 3.8) in recognition of the fact that Krugman's analysis was for Atlanta – the city that had the greatest increase in per capita congestion costs over the period considered. We then reduce the estimated national congestion costs of $78 billion per year by 50 percent to reflect a more reasonable estimate of the value of travel time. The estimated external cost savings from telecommuting under these assumptions are:

(30% possible telecommuters) · (average 20% reduction in commuting) ·
(2 peak multiple) · ($39 billion/year congestion costs) = $4.7 billion per year.

Summing up, the total estimated savings from telecommuting are $11.1 billion/year travel time for telecommuters, $7.3 billion/year for travel costs, and $4.7 billion per year for reduced (external) congestion costs, yielding total savings of over $23 billion per year. Assuming that these savings grow at a rate similar to the general growth rate we assume for the economy, these savings could be as much as $30 billion in ten years.[53] If half of this estimated effect of the increase in telecommuting could be attributed to the deployment of broadband, the net benefits attributable to broadband would be $15 billion per year.

3. Home Entertainment

The multi-channel video revolution of the 1980s and 1990s created enormous value for consumers. No longer confined to the limited diet of off-air broadcasting, consumers could choose from among scores or even hundreds of channels of movies, sports, news, travelogues, town meetings, and informational programs. As we show below, this explosion in choice created between $77 billion and $142 billion in annual value beyond the costs of the service (consumer surplus).[54]

The Napster sensation probably provides only a prelude to what is possible over household broadband connections. Downloading motion pictures or other video material, interactive television, interactive games, and even home editing of digitized entertainment material have not even begun in earnest. It is reasonable to assume that eventually the

---

[51] To project the civilian labor force in 2025, we multiplied the over-16 population estimate for 2025 times the current labor force participation rate. The Bureau of Labor Statistics does not project the civilian labor force past 2008.

[52] In accordance with the TTI report, we use an average trip length of 9 miles.

[53] In order to be conservative, we do not extend these projections past ten years.

[54] See Section VI.D, below.

contribution to consumer surplus of new video and other entertainment options created by widespread diffusion of broadband Internet access would be at least as great as that already created by cable television and direct broadcast satellites. This provides us with an estimate of $77 billion to $142 billion per year.

This estimate is consistent with the remarkable growth in home-entertainment spending by U.S. consumers since 1980, the year that cable began to grow as the result of FCC deregulation of cable signal carriage (see Table 3). Total expenditures have risen by about $56 billion since 1980 (nominal dollars). If the price elasticity of demand at current prices is substantially less than one, the consumer surplus from this increase could easily be more than $100 billion per year. We see no reason why the next broadband revolution – delivered by the Internet – should not be of equivalent value.

Table 3. Total Home Entertainment Spending, 1980–99

Year	Entertainment Expenditures by U.S. Households ($ Millions)	Year	Entertainment Expenditures by U.S. Households ($ Millions)
1980	4,657	1990	29,822
1981	6,292	1991	32,160
1982	8,199	1992	34,009
1983	10,374	1993	38,016
1984	12,742	1994	39,513
1985	14,708	1995	42,380
1986	16,915	1996	46,647
1987	19,869	1997	50,730
1988	23,250	1998	55,231
1989	23,685	1999	60,765

Source: Paul Kagan Associates, *Media Trends 2001,* at 44.

4. Home Health Care

Health care is the biggest industry in the United States economy. In 2000, health care expenditures were $1.3 trillion, or 13 percent of GDP.[55] Chronically ill patients, who represent 1 percent of the population, account for one-third of health care expenditures. Telemedicine could provide substantial cost savings in the monitoring of chronically ill patients, by substituting remote monitoring for in-person monitoring.[56] Remote monitoring can reduce the response time needed to address problems of chronically ill patients who are at home, resulting in fewer hospitalizations and emergency room visits. Because the health care industry is very competitive, we would expect almost all of the savings in cost to be passed on to consumers.

---

[55] Health care expenditures in 2000 were $1,299.5 billion (HCFA), and GDP was $9,825 billion (BEA) in current dollars. *See* HCFA, "National Health Expenditures Aggregate Amounts and Average Annual Percent Change, by Type of Expenditure: Selected Calendar Years 1960-2000", (downloaded from: *http://www.hcfa.gov/stats/nhe-oact/tables/t2.htm*) and Bureau of Economic Analysis, "GDP and Other Major NIPA Series", (downloaded from: *http://www.bea.doc.gov/bea/dn/gdplev.xls*).

[56] "Electronic House Call Knocks on Door of Telemedicine Market", *Medical Industry Today,* (June 19, 2000).

Telemedicine cost savings could not only reduce labor hours due in the delivery of home health care and in chronic patient management, but it could also reduce the incidence of emergency room visits and hospitalization, due to better disease management.[57] According to Cyber-Care, the manufacturer of Electronic HouseCall, an Internet-based system that allows caregivers to monitor patients in their homes, independent industry studies have shown that remote monitoring of patients' conditions can reduce costs by 35 to 40 percent, while also improving care.[58]

Assuming that chronic care patients are responsible for $400 billion in health care expenditures, the following table represents the potential cost savings from increased efficiency in chronic care management.

Table 4. Ultimate Cost Savings from Telemedicine (\$ Billions per Year)

Share of Households with Broadband Access	Improvement in Chronic Care/Disease Management				
	0%	5%	10%	15%	20%
10%	0	2	4	6	8
25%	0	5	10	15	20
50%	0	10	20	30	40
85%	0	17	34	51	68
100%	0	20	40	60	80

Despite the size of the health care industry in the United States, access to health care remains problematic in some regions, such as rural and inner city urban areas, and for less mobile populations, such as the elderly and disabled. Telemedicine would make health care more readily available to these populations, leading to an increase in consumer surplus. Already, a number of telehealth projects in the United States use high-speed data connections and videoconferencing technology to bring high-quality health care to rural and small-town residents.[59] In addition, the substantial savings in time associated with telemedicine would reduce the cost of health care for patients, thereby creating additional consumer surplus.

[57] Kaiser Permanente, the largest health maintenance organization in the United States, conducted a quasi-experimental study of telemedicine between May 1996 and October 1997, in which both the treatment and control groups received routine health care and the treatment group also received interactive remote health care in real time. On average, a traditional in-person visit took 45 minutes, whereas remote video visits required less than half as much time (18 minutes). Clearly, remote video visits allow health care professionals to see more patients in a day. In addition, even though the average direct cost of health care delivery was higher for the treatment group ($1,830 versus $1,167), the total mean cost of care excluding home health care was lower for the treatment group ($1,948 versus $2,674). Much of the lower healthcare costs were a result of lower hospitalization costs ($1,087 versus $1,940). Considering that some of the costs of remote health delivery are one-time costs (such as purchasing and setting up the equipment), cost savings would exceed the additional cost of remote home health care. Source: "Outcomes of the Kaiser Permanente Tele-Home Health Research Project," *Archives of Family Medicine* 9(1), (Jan. 2000).

[58] *Medical Industry Today*, *supra* note 56.

[59] *See*, for example, Jeff Tieman, "Dialing up High-Tech Medicine; L.A. University Opens Doors to its Third Telemedicine Center to Serve Urban Patients," *Modern Healthcare*, (Jan. 1, 2001) and "Telemedicine: Catching the Wave of the Future," *American Health Line* (Sept. 15, 1994).

Broadband access and reliable home networking will greatly facilitate the adoption of telemedicine. As David Schwartz, director of systems and networks for West Virginia University, Morgantown, and former director of the West Virginia Network, a state telemedicine initiative, points out, "the main concern with transmitting 3D images over a telecommunications network is the high consumption of bandwidth."[60] On the same point, Texas Telemedicine Project's Dr. Jane Preston says,

> And while local projects are making progress, telemedicine is not yet reaching into the homes because homes are still not linked to broadband networks. However, the technology is rapidly advancing and the price is dropping. When America's communications infrastructure evolves into wide access to two-way video, potential health care applications will include a full range of screening, diagnosis and the delivery of services, monitoring and treatment.[61]

Unfortunately, it is difficult to quantify the consumer surplus resulting from the greater availability of health care, so we do not attempt to do that here. Moreover, we have not included the potential benefits from telemedicine's contribution to improvements in the quality of health care in our quantitative estimates of the potential benefits from widespread adoption of broadband. We simply highlight the potential cost savings from telemedicine. For instance, if telemedicine results in a 10 percent decrease in health care costs and all the savings are passed on to consumers, the increase in consumer surplus would be $20 billion with 50 percent broadband penetration and $40 billion with universal service. These estimates assume that the demand for health care does not change when prices fall. If consumers demand more health care at lower prices, the increase in consumer surplus would obviously be greater.

5. Broadband Access and Telephone Services

Soon the demand for broadband will reflect not only the growing potential uses of the Internet, but also the prospect for using these broadband connections to obtain voice telephone services currently provided over a narrowband connection. The use of broadband access to carry voice – ordinary telephone calls – as well as data will convey substantial savings on consumers that are not captured in current demand estimates. Voice communications can be compressed, put in packets, and sent over an IP connection. If one has a network in the home or office, it is relatively easy to build a telephone that connects to the network and uses a Voice over Internet protocol (VoIP).[62] All one has to do to enable voice communications on a networked PC is to plug a headset into the audio ports on a computer and run Microsoft Netmeeting or AOL's instant messenger.

A broadband connection can support several voice connections – the exact number depends on the speed of the connection and the degree of compression of the voice signal. The current structure with two networks in the home (a voice and an IP network) and two connections to the outside world (a narrowband analog connection and a high-speed digital connection) appears inefficient. However, the transition to Internet telephony will take many years. The existing inside wiring, designed for voice systems, must be upgraded or replaced.

---

[60] "Telemedicine News: Starvision to Test Market Potential for Three-Dimensional Telemedicine Systems," *Health Data Network News,* (Mar. 20, 1998).

[61] *American Health Line, supra* note 59.

[62] *See,* for example, Cisco, "Technical Tips – Voice over IP (VoIP) Frame Relay (VoFR) and ATM (VoATM)" (downloaded from: *http://www.cisco.com/warp/public/788/voip/voip.shtml*).

The analog telephones must be replaced or connected to digital adapters, and there will be difficult issues of addressing, numbering, and directory services. Broadband connections must become highly reliable, and the quality of packet telephone calls must improve.

The cost savings from integrated access will be significant. Reliable Internet telephony would eliminate the need for second or third lines in households for teenagers or fax machines. The Federal Communications Commission (FCC) estimated that the average household spent $53 per month on local and long-distance telephone service in 2000 and there were 0.265 additional lines for each household with telephone service.[63] Within a few years, broadband access will permit consumers to substitute other services for these services that now cost $53 per month.

Part of that local telephone service cost is for the loop that is used for the broadband service. Consumers continue to incur most of those loop costs when broadband service is used, but they avoid the cost of the analog line card, the voice switch, and the voice transmission lines. IP telephony should lower the costs of both local and long-distance telephone service, while providing residences with the equivalent of several telephone lines. We estimate that such savings could average $25 per month per household. In addition, households with broadband service would get the equivalent of multiple voice telephone lines. We estimate that this additional service or option of service could be worth $5 per month to the average household.[64]

Thus, in the longer run (say a decade from now), broadband access could deliver voice communications benefits of about $30 per month, or $360 per year, to the average household with telephone service. If we assume that 122.2 million households have telephone service, these benefits would total $51.4 billion per year, assuming no growth in voice usage. The actual value could be much higher.

The substantial economic benefits (principally savings from expenditures on telephone service) created by providing multiple services over a high-speed line almost cover the cost of a high-speed line – we have estimated that benefits of $35 per month are created by a broadband connection that costs $40 per month. These savings are one reason that it is reasonable to expect that the fraction of households with high-speed access services will ultimately approach the fraction that has telephone service today.[65]

E. Summing Up the Consumer Benefits

Given the vast uncertainty about the future effects of broadband access, it is difficult to offer compelling quantitative forecasts of the economic benefits of speeding the deployment

---

[63] FCC, "Trends in Telephone Service", May 2002), (downloaded from http://www.fcc.gov/Bureaus/ Common_ Carrier/Reports/FCC-State_Link/IAD/trend200.pdf, Tables 3.2 and 8.4. Note that, because some households may have three or four telephone lines, this ratio has to be described carefully.

[64] The FCC's data indicate that the average household with telephone service had 1.265 access lines and paid local service fees of $35 per month in 2000. See FCC, *Trends in Telephone Service*, 2002, available at *http://www.bea.doc.gov/bea/dn/gdplev.xls*. Given second-line charges of between $7 and $10 per month and a price elasticity of demand of -0.35, the average consumer surplus for a second line may be estimated to be about $5 per month.

[65] Another way to look at it is to consider the engineering design. Why have two line cards at the telephone central office – one in the voice switch and one in the DSLAM – if the line card in the DSLAM has roughly ten times the capacity of the line card in the voice switch? If we were building the telephone network from scratch today, broadband access connections would be the natural building block.

of broadband access. Because broadband provides consumers and producers with a new alternative that has few harmful external effects that we can identify, it is reasonable to focus only on the benefits of broadband access.[66] These benefits are summarized in Table 5.

F. Producers' Surplus

Obviously, a successful innovation generates value for producers, as well as for consumers. For example, consumers have benefited handsomely from the development of automobiles and personal computers, but these two innovations left Henry Ford and Bill Gates among the wealthiest persons of their times. Their ability to produce automobiles or computer software and to sell them at prices above the cost of the inputs required in the production process provided them with substantial producer surplus. But others, such as skilled workers, owners of natural resources, and holders of complementary patents, also received some of this surplus.

Table 5. Summary of Consumer Benefits from Universal Broadband Deployment ($ Billions per Year)

Source	Low Estimate	High Estimate
Broadband Access Subscription (linear demand curve)	150	225
Household Computer and Network Equipment	20	54
Total Benefits	170	279
Estimate of Benefits Deriving from:		
Shopping	78	270
Entertainment	77	142
Commuting	15	15
Telephone Services	44	44
Telemedicine	40	40
Total Benefits	254	511

It is extremely difficult to estimate the potential flow of producer surplus from the widespread diffusion of broadband, but we can attempt to provide one based on various possible measures of the elasticity of supply of the services involved. The principal effects of greater broadband connectivity are registered in four different sectors of the economy:

- Telecommunications – the direct revenues from greater broadband connectivity.

66 We are not aware of any significant harmful externalities associated with the wider deployment of broadband access. Obviously, there are minor harmful externalities, such as the hydrocarbon emissions associated with the trips by installers to the residence and any externalities associated with the generation of electricity to operate the broadband equipment and computers. However, these systems use little power and the continuing externalities appear to be minor. If it turns out that broadband access will create significant harmful externalities, then our estimates will overstate the net benefits.

- The computer equipment and software industries – increased household purchases of equipment and software.
- The retail and wholesale sectors of the economy – greater efficiency in the distribution of consumer products.
- Entertainment – greater demand for entertainment products through the Internet.

We begin with our estimates of the increase in total sales to consumers of broadband services, computer equipment, general consumer goods, and new services that are developed because of widespread diffusion of broadband Internet access. We then calculate the share of such revenues that could be reasonably assumed to accrue to suppliers as producer surplus.

1. Broadband Services

The delivery of broadband services requires the use of existing telecommunications networks plus substantial expenditures on new electronics within such networks (multiplexers, line cards, etc.), as well as expenditures on consumer modems. Therefore, any increase in the demand for broadband will generate additional revenues not only for the suppliers of that service but also for the suppliers of the equipment used to deliver the service.

A substantial share of the revenues for modern electronics equipment accrues as rents to the intellectual capital required to develop it and to the substantial initial costs of building the facilities to produce it. It is obvious that today's price of a microprocessor or storage device does not reflect its marginal production costs, but it is difficult to estimate how much of any increase in demand will accrue as rents to the owners of the risky capital invested in the technology required for the production of such equipment.

For broadband services, we conservatively assume that 50 percent of the revenues are spent on the electronics and related equipment to deliver the service and that 20 percent of the remainder accrues as "quasi-rents" to the supplier of the communications service itself. We further assume that between 20 and 40 percent of the revenues absorbed by purchases of the requisite electronics equipment accrue as rents to the owners of scarce intellectual and physical capital required to produce it. Under these assumptions, we may calculate the producers' surplus for an expansion of broadband to 50 percent of households and to "universal service" proportions of 94 percent of households. The results range from \$6.2 billion to \$17.6 billion per year. (See Table 6 below.)

Table 6. Estimated Increases in Producer Surplus from Greater Broadband Connectivity ($ Billions per Year)

Source	Share of Revenues that Accrue as Rents to Producers		
	20%	**30%**	**40%**
At 50 % penetration			
Broadband Access Subscription	6.2	7.8	9.4
Telecom Services	5.0	6.3	7.5
Household Computer Equipment	10.6	15.9	21.2
Entertainment	12.8	19.2	25.6
At 94 % Penetration			
Broadband Access Subscription	11.7	14.7	17.6
Telecom Services	9.5	11.8	14.2
Household Computer Equipment	22.0	33.0	44.0
Entertainment	34.7	52.1	69.4

2. Computer Equipment

For computer equipment, we assume that the growth rate of spending would resume its 1991–95 growth rate of 14.3 percent at 50 percent broadband penetration. Were broadband to become universal, we assume a growth rate of 17.3 percent. These growth rates are much higher than the 10.4 percent growth rate that was experienced between 1995 and 1999.

To estimate the share of increased revenue that accrues to suppliers as producers' surplus, we estimated the percentage of revenues that will accrue as rents to intellectual capital. Similar to broadband service, a substantial share of the revenues for computer equipment accrues as rents to the intellectual capital used developing the technology. The price of a computer exceeds the marginal cost of building it. We examined the market-to-book ratios of major computer hardware and software producing firms (Dell, Gateway, Hewlett Packard, IBM, Microsoft, and Intel) in late 2002. The market-to-book ratios ranged from 0.83 for Gateway to 14.56 for Dell,[67] suggesting that the market value of rents above the acquisition cost of assets of as much as 93 percent. Given the variable costs of production, these rents are likely to be a substantially smaller fraction of revenues. Once again, we use a range of 20 to 40 percent of revenues as an estimate of the amount of additional revenues that are likely to accrue to producers as surplus. Under these assumptions, producers' surplus generated in the computer sector from widespread broadband deployment could range from $10.6 billion to $44 billion per year.

3. General Consumer Goods Distribution

Consumer goods distribution is a $1.5 trillion sector of the economy today. The efficiency of distribution will increase with the penetration of broadband, as more consumers move to online shopping. Because the consumer goods industry is a very competitive industry, we would expect the supply curve to be very elastic. This means that most of the

[67] Market-to-book ratios downloaded from Yahoo! Finance.

savings in cost will be passed on to consumers. The savings in distribution costs will largely be registered through a shift of from traditional distribution to on-line commerce. Traditional retailers will grow more slowly, and on-line retailers will grow more rapidly.

There likely will be substantial gains to the successful first movers in e-commerce, despite the current pessimism over the inability of some, such as Amazon.com, to achieve profitability. To achieve a reduction in total distribution costs of 3 to 10 percent, e-commerce will surely have to account for more than 10 percent of all retailing in 25 years. However, we have no basis for estimating how much of the substantial commerce involved will be appropriated as surplus by the successful innovators.

4. Entertainment

Entertainment spending by U.S. consumers has been growing consistently every year since 1980, the year the FCC deregulated cable signal carriage. We assume that entertainment spending with universal broadband service will continue to grow at the same rate that it did in 1980–99 (14.5 percent). With 50 percent broadband penetration, we assume that entertainment spending will grow at 7.5 percent.

The fraction of revenues that accrue to "talent" in the motion picture industry is substantial but very hard to estimate. Assuming rents of 20 to 40 percent of revenues, the increase in producers' surplus from an increase in entertainment spending ranges from $12.8 to $69.4 billion (see Table 6).

G. The Benefit of Faster Adoption of Broadband

The analysis in the preceding section offers estimates for the annual benefits from the universal or near-universal adoption of broadband access. The estimates do not account for the time required to reach this universality of broadband. Rather, they simply reflect the potential benefits available when broadband reaches maturity and perhaps becomes as ubiquitous as today's ordinary telephone service.

The value of these ultimate benefits *today* is lower because they are achieved gradually over time. The proper way to estimate such benefits is to calculate a time stream of benefits and then to discount those benefits back to the present. Figure 5 below illustrates three possible scenarios for the adoption of broadband access over the next 25 years. The graph shows the fraction of households with broadband access. Each scenario begins with 8 percent of households having broadband access at the beginning of 2001 and ends with 94 percent having broadband access in 2025.

Figure 5. Adoption of Broadband Access

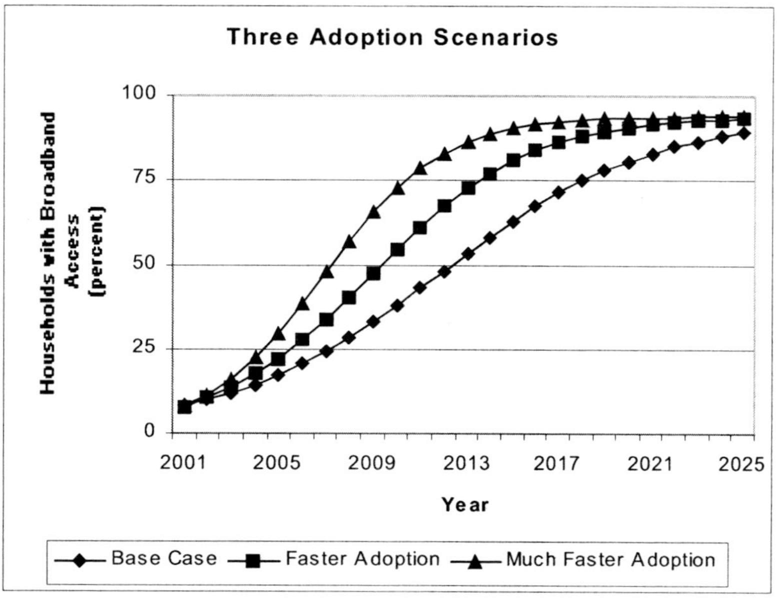

The lowest of the three curves is for a pessimistic "base case" scenario, which represents a relatively smooth transition from today's levels of broadband deployment to a level comparable to that of telephone subscription over a period of 25 years. Although the base case scenario is an S-shaped adoption curve, the curvature is gentle. In this scenario, the peak growth rate occurs in 2011 and is twice the average growth rate over the period. The middle curve represents a faster adoption scenario in which the growth rates are higher in the early years and peak growth occurs in 2009. The top curve represents a very rapid adoption scenario in which growth rates peak in 2007 at 11 percent of households per year.

Speeding up the adoption of broadband access provides benefits earlier. The present value of the difference between the base adoption scenario and the much faster adoption scenario of our example above is 140 percent of one year's worth of the benefits of ubiquitous broadband adoption by households.[68] Thus, if one assumed that, when fully adopted, broadband generated benefits of $300 billion per year to American consumers, a policy change that moved our society from the baseline adoption curve to the much faster curve in the Figure 5 would generate benefits with a net present value of about $420 billion.[69] The increase in the present value of producers' surplus would be about $80 billion. This acceleration is therefore worth $500 billion to U.S. consumers and producers.

A skeptic, on reading this, will necessarily have doubts – how could speeding up the adoption of a technology have such massive benefits? The key lies in the substantial benefits that ubiquitous broadband can convey to consumers. Once virtually everyone has the service, the network effects from developing new services become very large. Moving these benefits

[68] This was calculated using a discount rate of 10 percent and assuming a 2 percent per year growth in the economy.

[69] These present values are 2.8 and 4.2 times the ultimate value of broadband adoption when evaluated at an interest rate of 10 percent per year.

forward a few years can create very large benefits – even when evaluated from today's perspective.

Obviously, if these assumptions prove to be incorrect because some other innovation intercedes in the next 25 years, these benefits will prove to have been overestimated.

VI. CONCLUSIONS

Although it will take many years, the widespread adoption of broadband access service will bring enormous economic benefits to our economy. No doubt many of the impacts cannot be foreseen. But some benefits can. We have calculated estimates of the economic benefits using two quite different methods. The first approach uses a prospective demand function for high-speed access and then calculates the consumer surplus associated with that demand function with high-speed access priced at $40 per month. We also include the benefits to consumers of any higher-quality network and computing equipment that they would purchase to use with their high-speed network access and the benefits to producers of greater purchases of network services and computer equipment.

Our second approach attempts to identify the sources of the specific benefits that broadband access can provide and to calculate the consumer surplus associated with such benefits. For example, if broadband access means that a consumer no longer needs to buy a second telephone line at $20 per month for dial-up access, we would count that as a $20 per month benefit of high-speed access.

From these two approaches, we conclude that the annual consumer benefits from broadband could eventually reach three hundred billion dollars per year. In addition, producers will also benefit from increased demand for electronic equipment used in the delivery of broadband service, increased spending on household computer and networking equipment, increased spending on household entertainment, and improvements in health care delivery. These benefits could easily amount to another $100 billion per year if broadband becomes ubiquitous. Thus, a reasonable figure for the total annual benefits to the U.S. economy of the widespread adoption of broadband access in all its forms – ADSL, cable modems, satellites, 3G wireless, and others – is $400 billion per year.

A faster rollout of high-speed access services gives us these benefits earlier. Under optimistic – but still reasonable – scenarios the net present value of a faster rollout of high-speed access could be as high as $700 billion, and a mid-range estimate of the value of faster rollout is $500 billion.[70]

---

[70] We calculated the net present value of the difference between various adoption scenarios. With a discount rate of 8 percent the NPV of the difference between the base scenario and the faster scenario shown in Figure 5 is 100% of the end-year benefits and the NPV of the difference between the base scenario and the much faster scenario is 180% of the end-year benefits.

APPENDIX 1: ALTERNATIVE DELIVERY SYSTEMS FOR BROADBAND SERVICE

A. ADSL

ADSL is a simple concept. Equipment much like an ordinary modem can transmit data over the copper wire running from the home or office to the telephone company central office. Such digital subscriber line (DSL) modems can transmit data at far higher rates than is possible over voice-grade connections because they can take advantage of capacity in the copper wire that is not used for voice communications. Often the term *asymmetric digital subscriber line* (ADSL) is used. ADSL refers to a DSL system in which the transmission capacities in the two different directions are different – a split that can be useful for web browsing or distribution of audio or video programming.

ADSL service is limited by loop quality, distance, and the use of digital loop carrier systems. ADSL connections can be restricted or even made unworkable if the copper loop is impaired. For example, loading coils, which improve voice transmission on longer loops, must be removed for ADSL to work. The ability of copper telephone loops to carry high-speed data signals declines with distance. Thus, a copper loop a mile long might be able to carry ADSL signals at one million bits per second. An essentially identical loop, but two miles long, might be able to carry ADSL signals at only 300 thousand bits per second.

For decades, local exchange carriers have used a technology called digital loop carrier (DLC) to lower the cost of providing telephone service. In digital loop carrier systems, a high-capacity digital transmission line is run to a remote terminal in the neighborhood. At the remote terminal, the digital voice signal is converted to analog and transferred to copper loops. Digital loop carriers save money by substituting one high-capacity transmission system for many lower-capacity copper loops. Unfortunately, DLC systems do not provide a copper connection all the way from the telephone company central office to the subscriber location. The solution to this is to upgrade the capabilities of DLC systems to enable then to support ADSL service over the copper connection to the subscriber premises.

B. Cable Modems

Cable television systems can provide both telephone service and high-speed, two-way data communications service. In order for a cable system to be able to provide high-speed data service, the cable system must be modified to support communications on the return link, and a specialized two-way communications terminal must be installed in the home. Such two-way terminals are usually called cable modems, and the service is often called cable modem service.[71] Cable modem service provides data communications to users at data rates up to a few million bits per second.

Cable modem service has been relatively successful in the marketplace. All data sources agree that cable modem service is the most widely used form of high-speed Internet access. For example, the Yankee Group estimated that there were 3.7 million cable modem

---

[71] For a description of the development of the cable modem standard see CableLabs' Project Primer at http://www.cablemodem.com/docsisprimer.html.

households in the United States at year-end 2000. In contrast, the Yankee Group estimated that there were 1.7 million DSL subscribers.[72]

C. Wireless

The term *wireless* encompasses a wide range of possible technical alternatives, including satellite systems, cellular and PCS services, services in the MMDS and LMDS bands, and services in the unlicensed bands. Unlike DSL and cable modems, most of these technologies will require several years to develop and even longer to gain widespread use. They are also less familiar to many.

Satellite systems are the most widely used of the high-speed wireless options today. Two firms, DirecPC and StarBand, provide two-way satellite-based Internet access. These systems provide service at data rates of about 400 to 500 kilobits per second. Satellite firms appear to be targeting customers in rural areas who will not be offered DSL or cable-modem service. The next generation of satellite systems, due in orbit in about 2002, will offer higher data rates and much more system capacity.

Today, there are about 115 million subscribers to wireless telephone – cellular and PCS – service. Many of these wireless phones can provide Internet connections at about 10 kilobits per second – considerably slower than even today's dial-up connection. Typically, current-generation wireless phones do not provide always-on connections. This will change over the next few years as third-generation wireless systems, sometimes called "3G wireless," replace and supplement the current second-generation systems. The 3G systems promise to provide access at rates in the megabit-per-second range and will have the advantage of portability and lower cost than current systems or 2.5G systems.

The FCC has licensed two radio services, MMDS and LMDS, which also can be used to provide wireless Internet access to fixed locations. MMDS (multi-channel multi-point distribution service) is a radio service that was originally used to provide a wireless alternative to cable television. Sprint and WorldCom have obtained the right to use many of the MMDS licenses and are using some of their licenses to provide high-speed Internet access. MMDS technology can serve customers within a range of about 15 miles of a base station and requires a line-of-sight path between the antenna at the residence and the antenna at the MMDS service provider's base station. The FCC has also licensed a second service, the local multipoint distribution service (LMDS). LMDS operates at frequencies ten times higher than those used by MMDS. Consequently, LMDS transmissions are strongly attenuated by rainfall

Unlicensed radio bands provide yet another alternative for high-speed Internet access. The FCC has made several radio bands available for operation on an unlicensed basis using low-power radios.. A few hundred smaller ISPs have used unlicensed wireless to provide links to their customers. Typically, such systems run at data rates of about one million bits per second and can serve customers at ranges of up to about 15 miles.[73] It is still unclear how successful unlicensed radio services can be for small business and residential Internet access.

---

[72] Michael Goodman, "Residential Broadband: Cable modems and DSL", *The Yankee Group, Boston*, (Mar. 2001).

[73] Several ISPs that use unlicensed radio channels for access are listed at *http://www.cmc.com/lars/engineer/wireless/w-isp-list.htm*

In rural areas, where the probability of interference is low, unlicensed radio may become a real competitor.

Short-range communications using wireless LAN technologies are much less likely to suffer from harmful interference than longer-range systems. There has been substantial interest in providing broadband Internet access at public points, such as airports and restaurants using such technologies. For example, Starbucks has announced a program to put 11-million bit/second wireless LANs in all three thousand of their U.S. retail outlets.[74] Several airlines now provide wireless LAN connections in their airport lounges. These short-range services will probably not be used to provide access for homes or businesses.[75] However, they show the value that people attach to high-speed Internet access, and they will stimulate the adoption of high-speed Internet access more broadly.

APPENDIX 2: EXAMPLES OF PAST PESSIMISM IN NEW TECHNOLOGIES

A. Cable Television

To estimate future demand for broadband service without knowing what it may bring in the future would be much like trying to predict the demand for cable television service in 1974 when the only viewing alternatives that cable could provide were distant broadcast stations. At that time, pessimistic economists were predicting that cable would only spread to 30 to 40 percent of households in major markets. One of the current authors suggested that these predictions were decidedly too low, suggesting that 65 percent cable penetration was even possible – almost precisely today's level.[76]

B. The Photocopier

Chester Carlson developed a process for dry photocopying. He searched for years for a corporation that would commercialize his invention approaching, among others, General Electric, RCA, IBM, and Remington Rand. IBM hired the technology consulting firm Arthur D. Little (ADL) to assess the technology. After receiving ADL's pessimistic report about the likely demand for copiers, IBM chose not to pursue the technology further. A small firm, Haloid Corporation, subsequently decided to license the technology. After the product began to take off, the firm changed its name to Xerox.

---

[74] See Bob Brewin "Starbucks Takes Wireless Leap," *ComputerWorld,* (Jan. 8, 2001).

[75] Some people envision a world in which many such short-range wireless LANs, by cooperatively sharing capacity and relaying messages to a backbone connection, could provide substantial Internet connectivity. See http://www.elektrosmog.nu/.

[76] Robert W. Crandall and Lionel L. Fray, "A Reexamination of the Prophecy of Doom for Cable Television," *The Bell Journal of Economics and Management Science,* (Spring 1974).

C. Wireless Telephony

Many groups vastly underestimated the rate at which consumers would adopt wireless (cellular or PCS) telephone service. For example, a group of researchers at Cornell, in an excellent research effort, attempted to forecast the impacts of the yet-to-be deployed cellular technology.[77] Their analysis considered three scenarios of likely mobile telephone service penetration. Their medium scenario envisioned only 500,000 to 1 million mobile telephone subscribers over a 15 to 20 year horizon, and their most optimistic scenario envisioned no more than 10 million subscribers. By January 2000, actual subscribers were 86 million -- even their most optimistic scenario was exceeded by a factor of almost 9.[78] The Cornell authors were not alone. The forecasts by all parties in the FCC rulemakings for the cellular service fell far short of actual demand.

D. Computer Communications

The pioneers of packet communications and the Internet faced great skepticism from the established communications industry. In 1978, Larry Roberts, one of the pioneers of the Arpanet/Internet, wrote, "AT&T and its research organization, Bell Laboratories, have never to my knowledge published any research on packet switching."[79] He also described how ARPA, in an attempt to commercialize the Arpanet, approached AT&T about taking over the Arpanet, but AT&T declined the opportunity. The Arpanet ultimately became today's Internet.

APPENDIX 3: THE BENEFITS FROM EARLIER INNOVATIONS IN NETWORK INDUSTRIES

Broadband communications is hardly the first example of a revolutionary network technology. The development of railroads, electricity, the interstate highway system, the airline industry, energy pipelines, the telephone, and multi-channel video distribution (cable and satellite) are obvious examples of forerunners to the Internet and broadband. Each of these new technologies made major contributions in terms of economic value and wealth and in most cases the magnitude of their influence and the variety of their use far exceeded initial expectations at the time of their introduction. In this section, we review the published estimates of the benefits of some of these revolutionary technologies.

---

[77] Raymond Bowers, et.al., (eds.), Communications for a Mobile Socirty: An Assessment of New Technology. Beverly Hills, CA: Sage Publications, 1978, pp. 386-89.

[78] *See* CTIA, "Frequently Asked Questions and Fast Facts" (downloaded from: *http://www.wow-com.com/consumer /faq/articles.cfm?ID=101*).

[79] Larry Roberts, "The Evolution of Packet Switching," *Proceedings of the IEEE,* Vol. 66, No. 11, (Nov. 1978), 1307–1313.

A. Railroads

Early research on the effect of railroads on economic growth found that the railroad was the most important innovation of the last two-thirds of the nineteenth century. For example, Jenks points out that the railroad

> [a]ppears as the *sine qua non* of America economic growth, the prime force behind the westward movement of agriculture, the rise of the corporation, the rapid growth of modern manufacturing industry, the regional location of industry, the pattern of urbanization, and the structure of interregional trade.[80]

Cootner also maintains that the railroad was a crucial factor in American economic growth:

> Without the cheap land transportation that railroads provided, the United States – and more particularly the trans-Appalachian region – would have found it more difficult to use the economies of specialization that were so important to both the level of per capita income and its rate of growth in the nineteenth century.[81]

Cootner claims that the railroad expansion was due to several factors: (1) demand for coal driven by urbanization had increased, (2) water canals did not reach the higher altitudes where the coal mines were located, (3) passenger traffic between cities required faster transportation than the water canals could provide, and (4) cotton railroads were put in place where there was no other alternative.

Nevertheless, Fogel's pathbreaking research finds that the effect of railroads was not as great as previously thought. There is considerable evidence that shows a strong association between the diffusion of the railroad and economic growth. The causal relationship between railroad diffusion and economic growth, however, is predicated on the assumption that railroad transportation was cheaper than the alternative, water transportation. In the absence of water transportation (when the only alternative to railroad transportation would be land transportation by wagon), the social saving from railroads would be $4 billion per year, or more than one-third of gross national product in 1890. However, with water transportation as an alternative, the social saving was only $70 million per year, or less than 1 percent of gross national product in 1890. Thus, the development of canals and improvements to other inland waterways, in combination with the huge investments in railroads in the nineteenth century, were responsible for a very large share of U.S. GDP.

B. Electricity

The shift from steam to electric power took place between 1880 and 1930 in the United States. The use of electricity reduced the cost of energy, but more important, increased the

---

80 As quoted by Robert W. Fogel, "A Quantitative Approach to the Study of Railroads in American Economic Growth: A Report of Some Preliminary Findings," *Journal of Economic History*, (22)2, (June 1962), at 164.

81 Paul H. Cootner, "The Role of Railroads in United States Economic Growth," *Journal of Economic History*, 23(4), (Dec. 1963), at 477.

productivity of labor and capital input. Devine summarizes the advantages of electric power over other forms of power:

> The form value of electricity was due to the precision in space, in time, and in scale with which energy in this particular form could be transferred. Motors could convert electrical energy to mechanical energy precisely where the conversion was needed – the drive shaft of a machine. This conversion and transfer of energy could be exactly controlled with respect to time – that is, it could be started, stopped, or varied in rate as needed. And finally, electric motors could be accurately matched to the power requirements of machines. Thus, electric unit drive was an extremely flexible technique for driving machinery; and because of this flexibility, manufacturers could turn their attention away from problems of power production and distribution and toward improving the overall efficiency of their operations.[82]

Electricity was an invention that truly transformed manufacturing. The flexibility of electric power allowed for improvements in factory organization; machinery could now be arranged according to the natural sequence of manufacturing operations, rather than close to the source of power. Labor and capital productivity rose as electricity replaced steam as the source of motive power. In 1919, electricity accounted for 53 percent of the country's motive power and in 1929, 78 percent. During the same period, output per unit of labor rose from 186.6 to 321.5, and output per unit of capital rose from 48.9 to 74.5.[83] Perhaps not all of this increase in productivity is attributable to the switch to electric power. Anecdotal evidence, in the form of manufacturers' reports, however, suggests that most of it is. Crocker summarizes a number of these reports:

> It is found that the output of manufacturing establishments is materially increased in most cases by the use of electric driving. It is often found that this gain actually amounts to 20 or 30 percent or even more, with the same floor space, machinery, and number of workmen.[84]

C. The Telephone

As John R. Pierce of the California Institute of Technology points out, when Bell invented the telephone in 1876, society appeared to have been doing very well without it. Quickly, however, "the telephone has become more than a luxury or a convenience; it has become a basic part of man's world."[85] Colin Cherry, Professor of Telecommunication at the Imperial College, London, credits the development of the telephone system with having facilitated "revolutionary" progress in both the economic and social spheres of society.[86] From an economic standpoint, "telephone service is essentially organizational in function; it creates *productive* traffic"[87] by its very existence. According to Cherry, the importance of telephone networks lies in their "contribution to the organized bureaucracy that is the

---

[82] Warren D. Devine, Jr., "From Shafts to Wires: Historical Perspective on Electrification," *Journal of Economic History*, 43(2), June 1983, at 371.

[83] *Id.*, at Table 2.

[84] Crocker, *Electric Distribution of Power,* as cited by Devine, *supra* note 82, at 364.

[85] John R. Pierce, "The Telephone and Society in the Past 100 Years," in Ithiel de Sola Pool, ed., "The Social Impact of the Telephone", MIT Press, (1977), at 159.

[86] Colin Cherry, "The Telephone System - Creator of Mobility and Social Change," in Ithiel de Sola Pool, ed., "The Social Impact of the Telephone", MIT Press, (1977), at 113.

[87] *Id.*, at 114.

hallmark of modern industrial society."[88] The introduction of telephone networks changed the way businesses were organized. It was no longer necessary to be at the same location to conduct business. Weiman suggests an affinity between long-distance telephone service and two industries that rely on communication between two distant points: wholesale trade and hotels.

The largest users of toll and long-distance service in the mid-1920s were hotels, which furnished business executives and sales agents with lodging and vital services during their routine trips to the nation's economic capital. The second set of establishments included the corporate offices, often headquarter facilities, of national and multi-national firms. In both cases, toll connections were essential to conduct wholesale trade, whether to reach retail customers in the trade area or to keep close contact with distant production facilities and distribution centers.[89]

In two separate articles published in *Telecommunications Policy,* Cronin *et al.* show the effect of telecommunications infrastructure on the growth of the economy and productivity growth. Both articles analyze data from 1958 to 1990. Cronin concludes that the causal relationship runs both ways: investment in telecommunication infrastructure stimulated economic growth, and economic growth stimulated investment in telecommunications infrastructure:

> [A] feedback process [exists] in which telecommunications investment enhances economic activity and growth while economic activity and growth stimulate demands for telecommunications investment.[90]

Productivity growth in the telecommunications sector has been far higher than average productivity growth, and investment in telecommunications infrastructure has contributed to overall productivity growth.

> Without advances in telecommunications production, the U.S. economy would have experienced greater declines during the 1970s and a slower recovery during the 1980s.[91]

The authors also report that over the time period between 1975–90, the percentage of productivity gains in the economy due to telecommunications advances ranged from 18 to 37 percent, averaging 25 percent.

D. Multi-channel Video Distribution

Until the 1980s, most U.S. households were limited to three or four television viewing options because of the FCC's spectrum allocation policy. Some communities had even fewer

---

[88] *Id.*, at 113.

[89] David F. Weiman, "Building Universal Service in the Early Bell System: The Reciprocal Development of Regional Urban Systems and Long Distance Telephone Networks," mimeo, Russell Sage Foundation, (2001).

[90] Francis J. Cronin, Edwin B. Parker, Elizabeth K. Colleran, and Mark A. Gould, "Telecommunications Infrastructure and Economic Growth: An Analysis of Causality," *Telecommunications Policy,* 15(6), (Dec. 1991), at 533.

[91] Francis J. Cronin, Elizabeth K. Colleran, Paul L. Herbert, and Steven Lewitzky, "Telecommunications and Growth: The Contribution of Telecommunications Infrastructure Investment to Aggregate and Sectoral Productivity," *Telecommunications Policy,* 17(9), (Dec. 1993), at 679.

viewing options because of their remote location or because of topographical obstructions. To relieve the latter problem, small "cable" television systems began to retransmit signals from nearby locations in the 1950s. In time, cable television evolved into a delivery system with enormous capacity, but only after nearly two decades of struggling against restrictive FCC regulation.

The most restrictive of the FCC cable rules were relaxed in 1979, a year in which there were only 15 million cable television households in the United States. Twenty years later, the number of cable subscribers had expanded to more than 65 million.[92] Moreover, in the 1990s, two new high-powered satellite broadcast services were launched. These two services, DirecTV and Echo Star, each offer more than 200 channels of programming and now have more than 15 million subscribers between them.[93] To counter this new form of competition, cable television operators have adopted digital transmission techniques to increase their capacity to 100 or more channels.

This remarkable expansion in television service has created substantial value directly and indirectly for consumers. Hundreds of new program services have been developed and placed on satellite transponders to feed the thousands of local cable systems. These services are available to the satellite broadcasters, as well. More than 80 percent of all television households subscribe to cable, a satellite service, or both. The value of the packages they can now receive is many times the value of the three to four off-air channels that were available to generations ago.

In a recent book, one of the current authors and Harold Furchtgott-Roth estimate that in 1993, an additional cable channel was worth \$0.70 per month to the average household, and an additional broadcast channel was worth \$0.93 per month, all expressed in 1992 dollars.[94] Because the average household was offered 23 basic cable channels and 6 off-air stations at the time, one may conclude that the 30th channel choice would be worth \$0.70 per month, while the 7th choice would be worth \$0.93 per month. For purposes of extrapolating to hundreds of channels, therefore, we assume that this rate of decline holds as we move out from the 30th channel. Quadrupling the number of choices from 30 to 120 would reduce the marginal value of a channel by about 25 percent to about \$0.53 per month in 1992 dollars.

How much is the option to receive up to 200 channels worth to households today, given the above results? Assuming that households could receive 6 channels off the air today without cable or satellite services, the next 94 channels would be worth \$61.43 per month. For all 105 million U.S. households, these 94 channels are worth \$77.4 billion per year. An additional 100 channels would increase this total value to \$142.1 billion per year. Thus, the development of these two new technologies that offer as many as 200 channels per year directly create nearly \$150 billion in value for consumers every year, but they also create value indirectly. By expanding the network of opportunities for exploiting the value of new entertainment, information, and interactive services, cable and satellite broadcasting have induced the production of new products that would never have been possible had cable simply

---

[92] U.S. Bureau of the Census, "Statistical Abstract of the United States" (downloaded from: *http://www.census.gov/prod/2001pubs/statab/sec18.pdf*), at 567.

[93] Paul Kagan Associates estimates 15.3 million subscribers by year-end 2000. *See* Paul Kagan Associates, "Economics of Basic Cable Networks 2001," (July 2000), at 106.

[94] Robert W. Crandall and Harold Furchtgott-Roth, "Cable TV: Regulation or Competition," Brookings Institution Press, (1996), at Table 3.2.

remained a limited retransmission service with less than 10 million subscribers, as it was for nearly two decades.

The proliferation of cable and satellite has clearly created new video production opportunities that are exploited on other media, such as videocassettes, foreign television, and movie theaters. In addition, widespread cable and satellite coverage of sports leagues have created new markets for such sports as soccer, lacrosse, tennis, horse racing, and a variety of regional college athletics.

Finally, the growth of the multi-channel distribution systems has created huge rents for the inputs to the *production* of cable-satellite programming. This value is not captured in the consumer welfare calculations above. These rents have become so large that the entire Los Angeles MSA's economy can be put at risk by the threat of a writers' strike.

Chapter 9

BROADBAND DEPLOYMENT: TOWARD A MORE FULLY INTEGRATED POLICY PERSPECTIVE*

Johannes M. Bauer, Junghyun Kim and Steven S. Wildman
Department of Telecommunication
Michigan State University

1. INTRODUCTION

Broadband access to the Internet facilitates a wide range of services and applications that are difficult to configure or inconvenient to use in a narrowband environment. Viewed as a critical component of the next generation information infrastructure, governments around the world have made promotion of broadband diffusion a critical goal of communication policy.

Recent data from the Federal Communications Commission (FCC, 2002) indicate that broadband services are broadly available within the United States. As of June, 2002, subscriptions to high speed services were reported for 84 percent of US ZIP codes containing 98 percent of the US population, and about 16 percent of households subscribed to a broadband service.[1] Approximately 18 percent of ZIP codes have but one service available, while there were 10 or more service options in about 11 percent of ZIP codes (located in densely populated urban areas). While a number of technologies may be used to provide broadband service, consumer subscriptions in the US are dominated by cable modem services and digital subscriber lines (DSL), which account for just under two-thirds and just over one-third of consumer and small business subscriptions, respectively.

In the United States the diffusion rate for broadband is well within the range of growth rates experienced in their early years by other information technologies. (Pepper (2002) Broadband's diffusion rate has been higher than that of color television, VCRs, mobile phones, and CD players, but slightly below the initial diffusion rate for PCs and significantly

* The authors want to thank Ping Gai for his contributions to the background research supporting the analysis presented in this chapter.

[1] The 84 percent figure is an imperfect gauge of availability because broadband is not necessarily available to everyone in those ZIP codes reporting subscriptions and because the number of service options varies widely among locations.

below that of black and white television. Broadband proponents argue that a multitude of benefits are contingent on widespread broadband availability. Thus, despite apparently favorable growth comparisons with other technologies, there persists a widely articulated belief in the US that broadband penetration is not growing fast enough. While the US ranks fairly high in international comparisons, the fact that broadband penetration has grown much more rapidly in a few other countries, most notably Taiwan, Hong Kong, and Korea, with 31 percent, 36 percent, and 43 percent household penetration in 2002, respectively (ITU, 2003), is sometimes used to bolster this argument.[2]

In the United States, the federal government and most state governments have developed programs to promote more rapid adoption of broadband services, and there are numerous local initiatives as well. Other industrialized nations also search for frameworks for broadband policy although there is a wide range of views as to what the most effective policy framework should look like. Policy experiments range from laissez-faire approaches intended to free entrepreneurs from regulations detrimental to broadband investment to direct government ownership and operation of networks. Given the relatively high degree of uncertainty surrounding broadband, such diversity may be desirable as it can promote the generation of social knowledge as to what models work and which ones do not. Nevertheless, a more thorough conceptual understanding of the role of public policy should help elevate the debate.

Not apparent in the comparisons of the growth of broadband penetration with growth rates for other communication technologies, and generally overlooked in the broader debate over broadband policy, is the fact that broadband has a number of attributes that distinguish it from other communication technologies in ways that influence both the economics of diffusion and the nature of the benefits it might provide.

- Broadband refers to a level, or quality of service, that may be provided by a number of very different technologies.
- Broadband is a general purpose technology (or set of technologies), with a wide variety of potential applications, many of which are likely to appeal to only small subsets of potential users.
- Components of the distribution infrastructure are shared among users at levels of geographic aggregation ranging from the local neighborhood to the national and international.
- Bandwidth requirements vary dramatically among applications.

These characteristics of broadband may account for the otherwise remarkable feature of the policy debate over broadband—that it has been conducted without a generally accepted definition of broadband, let alone a systematic framework for evaluating policy options.

This paper is a first step towards a more comprehensive and integrated framework that explicitly accounts for the four distinguishing characteristics of broadband identified above. Section II is a brief review of the various positions put forth in the current debate over broadband policy. Section III provides a brief overview of broadband technology and the

---

[2] While international comparisons of this sort may be useful for developing a general sense of how penetration rates compare for different nations, it is important to note that these estimates are subject to considerable error and may vary widely among sources. For example, the ITU reports a 10 percent household penetration rate for the United States, compared to the FCC's recent estimate of 16 percent reported above. Because the ITU

architecture of broadband distribution systems. Section IV examines the economic implications of this infrastructure, focusing on the implications of shared costs for policy design in different local circumstances. Section V examines in more depth the nature of the benefits attributed to broadband and shows how they further complicate the design of appropriate policies. Section VI examines the spatial character of broadband benefits and the costs of certain broadband-enabled services and argues that these spatial relationships must be taken into account in formulating broadband policy. The policy implications of the framework and analyses developed in this chapter are briefly summarized in Section VII.

II. The Status Quo of Broadband Policy

Governments worldwide have adopted measures to support the rapid diffusion of broadband. With few exceptions such as Firth and Longstaff (2002) no serious challenge to the basic premise that broadband is desirable has been raised. There is widespread agreement that policy should encourage broadband deployment and reduce digital divides. The vast majority of the policy debate centers on the appropriate means to realize the potential benefits from broadband. Given the increased belief in market forces and competition, there is a clear bias to use competition as the main driving force and the role of the government is envisioned as organizing the market. Important roles are the establishment of a clear and conducive regulatory framework and perhaps measures to increase the awareness of the benefits of broadband. Not surprisingly, proposals in favor of more or less government seem often shaped by the particular ideological stance of a stakeholder. This section first reviews some of the pending policy initiatives. The next section develops an analytical framework as a broader context for this discussion.

Proponents of less government involvement support their view with five arguments. First, since broadband systems are in the early stage of deployment and development the government lacks sufficient information to design appropriate policy measures. Therefore, governments should wait until it is clear which services private investors will not provide and why. Second, for similar reasons it might be too soon for governments to define areas that are not commercially viable and thus need financial support. Third, the very initiatives that governments can adopt to boost competition can distort broadband markets. Such distortions can change the shape of competition in broadband markets and may even slow the development of the private sector's competitiveness. Fourth, governments' support for certain broadband providers or municipal networks might create local monopolies and have a negative impact on local industry innovation and efficiency. Last, where governments intend to provide financial support to public or private institutions, such as schools, libraries or hospitals, to help them build up broadband connections, it might be hard to evaluate and justify government support without reliable criteria for the investments (OECD 2002).

Table 1. Overview of federal, state and local broadband policy measures in the US

Type of Support	Federal	State	Local
Information	Centers of excellence		
Communications	Centers of excellence		
Indirect financial support	Support for R&D E-rate program	Tax credits (MI) Low interest loans (MI)	Link Michigan Community networks
Direct government ownership		Multi-Use Network (Colorado)	Ashland, Oregon

Proponents of more active government involvement emphasize, first, that market forces will be insufficient to bring advanced broadband services to certain rural or poor communities with limited profitability (APT 2002). Second, in a broadband environment, content providers may be dependent on downstream network service providers and government intervention may be needed to assure sufficiently open network access conditions. Direct government ownership is one alternative to create an open access network infrastructure. Third, government investment or subsidies can speed up the upgrade of a regional network where it might take several years to attract private broadband providers. Last, governments are believed to be responsible for acting in the public interest when there are no concrete consumer demands in the broadband market during its early stage of growth (OECD 2002).

Governments can become involved as owners and operators of broadband infrastructure, as is the case in several U.S. states (e.g., Maryland, Iowa) and at the community level. More often, the government designs measures to support private sector investment and use. Direct supply-side measures include direct procurement programs, tax incentives, subsidies and grants, or the formation of public-private partnerships as is the case in many community network projects. More indirect supply-side measures include general tax incentives for research and development in ICT. Government measures can also affect the demand side, for example, in the form of the U.S. federal E-rate program intended to assist schools and libraries to get access to fast internet connections.

In the U.S., most states and many communities pursue specific programs to facilitate the diffusion of broadband services (see table 2 for examples). One recent example at the state level is the Michigan Hi-Speed Internet Plan, enacted by the Michigan legislature in March 2002. This law established the Michigan Broadband Authority (MBA) as a non-profit organization to provide low-interest loans to communities and businesses in support of broadband networks. Another Michigan initiative is the Link Michigan program, which awards planning grants to groups of communities to explore options for higher speed network access. Moreover, a large number of communities have experimented since the 1980s with community networks. The experience with these networks is mixed, with some very successful projects and others that failed.[3]

[3] For a list of selected projects and hyperlinks to them see LinkMichgan, Regional Telecommunications Planning Program, Resource Book, available at <http://lm.michigan.org/commnetprojres.html>.

Table 2. Selected local broadband initiatives

	Objectives	Technology	Organization and financing
Electronic Villages, Virginia	Expansion of internet access outside of local university with broad synergies for political and economic life	Mix of Ethernet, fiber optical net, and mobile communications	Initially funded by communities and Virginia Tech, since 1995 private investors
Yolo Area Regional Network (YARN), California	Gateway to all local activities, including local politics, e-government, economy and education	Mix, final users predominantly DSL, some ISDN and	Privately owned, lower costs through demand aggregation
Eagle Mountain, Utah	Internet access for private and business users	Wireless communications	Privately owned and operated
Ashland, Oregon	Broadband internet access, cable television	Hybrid fiber co-ax network	Owned and operated by municipality
Prairie iNet (Iowa, Illinois, Kansas)	Broadband internet access	Wireless service in unlicensed 2.4 GHz band	Privately owned and operated
Multi-Use Network Project (MNT), Colorado	Access to efficient telecommunications services for rural communities and small and medium-size enterprises	Mix of several fixed technologies	Public-private partnership between State of Colorado and Qwest
Totally Web Government Program (TWG)	eGovernment	Mix of fixed technologies	National League of Cities, IBM, National Association of Counties

III. An Overview of Broadband Technology

Broadband refers to a set of general-purpose electronic communications technologies rather than one specific technological solution. Common features include significantly higher bandwidth than dial-up networks, the reliance on some form of packet switching, and "always on" functionality.[4] Broadband networks are also superior to traditional networks in terms of latency (the time it takes to deliver a packet across the network) and jitter (the variance in latency).

There is no common agreement as to what bandwidth threshold must be passed to have broadband. The U.S. Federal Communications Commission (FCC), as well as the Canadian Broadband Task Force, chose 200 kbps for downstream transmission as the minimum bandwidth for broadband service, although this is not a rigid number and may shift upward with changes in technology and uses (OECD 2002). The International Telecommunications Union (ITU) considers a transmission speed of 1.5 to 2 Mbps as broadband. The Computer Science and Telecommunications Board (CSTB) of the National Research Council (NRC) proposed a flexible definition capturing the dynamic evolution of broadband. The two-part

[4] The always-on feature is independent of bandwidth, but has come to be accepted as a defining characteristic of broadband.

approach defines broadband as transmission capacity in the local access link that is sufficient to not constrain a user's ability to run advanced applications. Moreover, broadband should provide sufficient performance and penetration of the potential user base to encourage the development of new services and applications (NRC 2002). The NRC approach suggests that as long as the computer technology supporting applications continues to advance, so must such a constraint-focused definition of broadband. Nevertheless, it is also true that for policies that must operate in both the near and longer terms, bright-line bandwidth criteria may be necessary to formulate policy goals in implementable terms and to gauge progress in achieving them.

Figure 1. Basic broadband network architecture

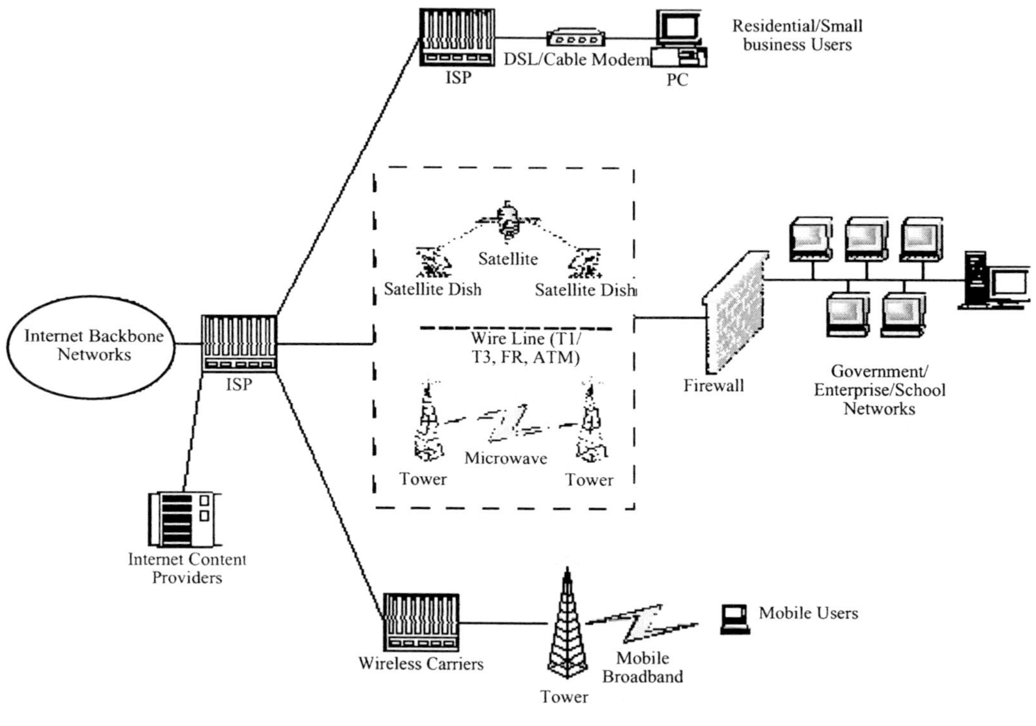

Some broadband access platforms are evolutions of existing networks whereas others require fully new networks (see Figure 1). The different types of Digital Subscriber Line (DSL), cable modems, and powerline communications are upgrades of already existing network infrastructures. Fixed wireless access (FWA) technology and broadband wireless solutions require more significant network investment. For example, IMT 2000 or 3G wireless services can only be offered after the installation of a new radio access network. The investment costs of these different solutions vary widely as does the scalability of the networks. For example, DSL and cable modem services require a smaller sunk investment than satellite-based services and their capacities can be adjusted more easily to varying demand. Broadband access technologies therefore offer a range of cost-functionality combinations. The technological and cost characteristics are more similar at higher levels of the network – the middle mile and backbone networks – which typically use similar topologies and protocols. Table 3 provides a summary overview of the spectrum of broadband access technologies and some of their key features.

Table 3. Features of broadband technologies

Types	Sub-types	Characteristics & features
DSL (Digital Subscriber Line)	ADSL (Asymmetric DSL)	-Asymmetric because of its focus on the downstream transmission of data to end users -Most common form of DSL for residential and small business users -Range from 1.5 to 6.1 Mbps in transmission speed
	G. Lite	-Slower ADSL that does not use line-splitting in the loop, but filters traffic at the telephone company's facilities -Up to 1.5 Mbps for downstream -From 32 Kbps to 512 Kbps for upstream
	HDSL (High DSL)	-Used in corporate networks and between telephone companies and end users -Equal bandwidth upstream and downstream -Up to 2.3 Mbps in transmission speed
	IDSL (ISDN DSL)	-Closer to ISDN speed, not broadband technology -Symmetrical
	RADSL (Rate-Adaptive DSL)	-A kind of ADSL technology that adjusts the access speed based on the condition of the line
	SDSL (Symmetric DSL)	-The same technology as HDSL with a single line -From 1.5Mbps to 2Mbps in transmission speed in each direction
	VDSL (Very High DSL)	-Higher data rates over shorter distances -From 51 to 55Mbps in transmission speed
Cable Modem		-Maximum speeds from 1Mbps to 10Mbps -Reduced transmission speeds if many users connect and transmit data at once
Fiber Optic Cable		-High transmission speeds of up to 10Gbps -High implementation costs
FWA (Fixed Wireless Access)		-Maximum speed of 150 Mbps in point-to-point systems -Maximum speed of10 Mbps in point-to-multipoint systems -Transmission speed falls if many users access simultaneously
Satellite		-Upstream transmission rates of 150-400 Mbps where upstream service available -Downstream transmission rates of 400-500 Mbps Service disruptions possible in bad weather
Mobile wireless		-IMT 2000 or 3G mobile -Up to 2Mbps data transmission
Power lines		-Electrical power networks -1-2Mbps transmission speeds

Broadband technologies are general purpose, shared networks. The general purpose nature of the technology means that individual users may use the network for multiple purposes such as voice, e-mail, online games, and various streaming services. The costs of significant portions of the physical and logical infrastructure of broadband services are also shared among users, as is evident in Figure 1. Compared to a circuit-switched voice network, sharing starts at nodes "closer" to the end user. The overall cost structure consists of a relatively high level of fixed compared to incremental costs. The high level of cost sharing among users and among services raises complex issues with respect to the apportionment of shared costs, which may influence the market structures of local broadband markets in ways that will vary by locality (Faulhaber and Hogendorn 2000). Policy options are also impacted, as is discussed in Section

IV. Shared Costs and Policy Choices

A. Basic Efficiency Framework with Purely Private Benefits

The physical components of a broadband infrastructure consist of the equipment employed by users to access the network and various elements of the network that connects them. If we ignore satellite services, it is convenient to think of the distribution links of a broadband infrastructure as potentially having four distinct components. At the core of the infrastructure are the national and international Internet backbone networks that are accessed through network-access-points (NAPs) in major metropolitan areas. Extending outward from the backbone NAPs are regional networks that connect more distant communities to the backbone. Communities are served by their own local distribution systems, which typically connect to individual users through feeder lines extending into neighborhoods. Dedicated plant connects a household to the neighborhood feeder system.

All but this last link are shared among users. If we ignore the controversy over whether such dedicated elements in telecommunications networks are properly treated as shared for cost allocation purposes,[5] we can think of the broadband infrastructure as having four levels of shared transmission elements and one that is not. Only for the dedicated facilities connecting users to neighborhood distribution systems are implementation decisions ultimately made by individual users. Implementation decisions at all other levels are, to greater or lesser degrees, matters of collective choice. How these decisions are made may depend greatly on local circumstances. This has important implications for the design of policies to promote broadband diffusion.

The remainder of this subsection presents a framework for classifying policies for broadband promotion in terms of how well they address policy challenges that may vary with local circumstances. We begin by examining the economics of broadband adoption for a single neighborhood and then extend the analysis to larger geographic aggregations of users. We start with the neighborhood because the challenges to policy design are more simply described at the neighborhood level, and because, as will become apparent, all the problems associated with cost sharing at the higher levels of aggregation may be present in the

[5] To the extent that users utilize broadband to communicate with each other, this last link is shared too. In a telephone network, this last link is the local loop, and whether the costs of a loop should be borne entirely by the user connected to it has been a matter of controversy. This debate is reviewed by Aron and Wildman (1996).

neighborhood as well. To simplify the analysis, it is assumed initially that there is a performance-based definition of broadband and a single technology for implementing it.

We will discuss policy options as they might apply to a hypothetical neighborhood with n potential broadband subscribers. Serving the neighborhood requires provisioning a common broadband pipe to which individual households are then connected via connecting lines that serve only them. While a mathematical formulation is not necessary to the arguments developed below, some additional clarity may be gained from representing certain relationships algebraically. Therefore, let K be the cost of the common pipe, or neighborhood distribution line (NDL), and let b_i be the benefit of broadband to the ith household net of the cost of provisioning a line connecting it to the NDL. Without loss of generality, we assume that $b_i \geq b_{i+1}$ for all $i \leq n$-1 and that household m is the last one for which the net benefits of connecting the household to the NDL are nonnegative. If there are no network externalities or other social spillover benefits not captured in the benefits of broadband as perceived by individual households, the goal of policy should be to promote adoption by all subscribers located in neighborhoods where the sum for all households of the net benefits of connection exceeds the cost of the NDL. The criteria developed below for assessing broadband policy options in the absence of network externalities and social spillover benefits will also be useful starting points for assessing policies when they are present. In discussing policies for promoting broadband penetration, we will distinguish between five different neighborhood situations and identify neighborhoods by the type of situation that characterizes them. Thus, for example, a neighborhood best described by situation 1 will be called a type 1 neighborhood.

Situation 1: *The sum of net benefits for broadband users in the neighborhood is less than the cost of the NDL.*

Algebraically, this situation is described by condition (1).

$$\sum_{i=1}^{m} b_i < K \tag{1}$$

With purely private benefits, an efficiency motivated broadband policy should not promote adoption in type 1 neighborhoods.

Situation 2: *The sum of the benefits to broadband users exceeds the cost of the NDL, but it is necessary to negotiate user fees that vary by customer to cover the NDL costs of a subscriber-supported service.*

Condition (2.a) is an obvious component of an algebraic description of situation 2.

$$\sum_{i=1}^{m} b_i \geq K \tag{2.a}$$

The possibility of a type 2 neighborhood is implied by a downward sloping demand curve for broadband service within a neighborhood. To see this, suppose that the cost of connecting a household to the NDL is a constant, c, for each household. Let q be the profit maximizing single price for broadband service and let household z be the marginal customer just willing to pay q for the service. Then $q=b_z+c$. It is entirely possible that $qb_z<K$, even though it would be possible to cover the cost of the NDL if the inframarginal consumers who value the service at more than q could be charged higher fees that vary with the benefits they would receive.

To illustrate this possibility, consider a neighborhood with 2 households for which K=$7 and c=$2. b_1=$5 and b_2=$3, so that $q=b_2+c$=$5. Charging both households $5 for broadband service would generate $10 in revenue, which is less than the total cost of $11. On the other hand, if household 1 paid $6.50 for service and household 2 paid $4.50, the $11 in total costs would be covered and each household would realize benefits of 50¢ over and above the price paid.

It might seem that $zb_z<K$ would be a second part of the algebraic characterization of situation 2, but this is not the case. Suppose that the neighborhood was such that $zb_z=K$. Then if any one of the households 1 through z refused to take the service, there would be no single price at which it would be financially viable. A household could use this threat to bargain for a lower price for itself, which would then have to be offset by higher prices paid by other subscribers. Thus, a single price strategy is sure to be viable only if a neighborhood has enough potential subscribers with a high enough willingness to pay that even if a coalition of those subscribers threatened to not take service, the remainder would be willing to pay a common price high enough to cover the cost of the NDL.

More generally, let g be the size of largest group that can effectively organize to negotiate a lower price for its members, let t be the subscription price that maximizes profits if broadband service is sold only to the members of the neighborhood not part of the group of g, and let v be the marginal subscriber outside the group of g for whom willingness to pay is just equal to t. Then the viability of a single price strategy may be threatened by strategic bargaining if condition 2.b is satisfied.

$$b_v(v-g)<K. \tag{2.b}$$

(2.a) and (2.b) constitute the algebraic description of situation 2.

To illustrate this possibility, we add subscribers 3 and 4 to the two subscriber neighborhood used above to show how benefits may exceed costs even if there is no single price at which a broadband service would be financially viable. $b_3=b_4$=$2, so now, for c=$2, q=$4. However, any one of the four members of the neighborhood can make a single price financially nonviable by refusing to take the service.

Situation 2 may present the most significant challenges for broadband policy to overcome. If $zb_z<K$, it will be necessary for some neighbors to pay more for service than others if the service is to be supported by subscriptions. The cost of the bargaining required to determine a set of varied, but financially sufficient, prices may itself be sufficient to render a commercial service non viable. Resentments over the apparent unfairness of some subscribers paying more than others may also be a barrier to implementation. Plus, some neighbors may try to reduce their own assessments by underreporting their true valuations of the service. As condition (2.b) shows, these problems may also be present due to coalition formation even if $zb_z>K$.

Commercial suppliers may also fear that subscribers will seek to renegotiate for lower, and ex ante unremunerative, prices once the NDL is in place, because a supplier's bargaining position is considerably weaker after sunk costs have been incurred than before, when refusal to build out a neighborhood is still an option. Plus, commercial suppliers may find it difficult to determine whether a neighborhood really has type 2 status. Potential subscribers' self-reports of willingness to pay for broadband may be overly optimistic, or deliberately inflated to increase the likelihood that broadband will be offered. There is also the risk that borderline neighborhoods may change status from situation 2 to situation 1 due to changes in the potential subscriber base after the neighborhood feeder line has been installed.

Policy may contribute to broadband adoption in type 2 neighborhoods in several ways. One is by facilitating negotiations over who pays how much when payments that vary among subscribers are necessary to support the service. This might be done, for example, by supporting the types of cooperative associations that first provided telephone service in many rural areas in the United States. Local governments and/or community organizations might also canvas potential users to certify demand and secure commitments from key customers in advance to take service at pre-specified prices to help reduce suppliers' investment risks.

Public investments in infrastructure may also help resolve coordination problems in situation 2 neighborhoods. If prospective subscribers' perceived benefits from broadband and local tax payments both vary with income or the assessed value of their real estate, funding all or part of infrastructure costs with public moneys may be a way to avoid the costs of haggling over contributions.[6] Suppose that a broadband levy included in property tax assessments generated payments equal to half the net value of service from each of the four members of our hypothetical neighborhood. That is, tax collections of $2.5, $1.5, $1, and $1 from residents 1, 2, 3, and 4 respectively. A common subscription price of $2.25 would then be sufficient to cover the remaining costs of the service. This might have the added benefit of extending service to potential subscribers who value the service less than do households 3 and 4—a household 5, for example, who is willing to pay only $3 for service. The downside of this approach is that there may be households who value the service less than the cost of connecting them to the NDL, but who still end up paying taxes to support it.[7]

The public contribution might be less direct than a broadband levy collected through taxes. Among prospective broadband customers, city and county governments might value broadband service the most, and could make commitments that would reduce the contributions to common costs that would have to be collected from other prospective subscribers. For example, suppose that subscriber 1 in our example is the municipal government rather than a household and it agrees to commit $4 of its $5 in net benefits toward the common cost of $7. Service could then be offered to the other four customers at $2.75 each. This could be accomplished through a government owned network, or through a government negotiated arrangement with a private provider.

Situation 3: A broadband service can be supported with a single per customer charge and there is no subscriber coalition large enough to pose a threat to this pricing strategy; but the level of demand in the neighborhood is not sufficient to support two competing broadband suppliers.

Condition (3.a) is one obvious component of the algebraic description of situation 3.

---

[6] Of course, price discrimination is also a mechanism that may be employed by a commercial broadband supplier for collecting different contributions to common costs from different customers. This can work and solve the coordination problem as long as subscribers are passive price takers.

[7] This discussion of the efficiency and coordination advantages of tax support for broadband infrastructure investments ignores the possibility that taxes may distort other consumption decisions.

$$b_v(v\text{-}g)>K. \tag{3.a}$$

(3.a) just says that there is no coalition of subscribers large enough to block implementation of a single price service. Define p(n) as the equilibrium price in a broadband market served by j firms and s(n) as the market share of the nth largest firm in the market,[8] which in this case is a neighborhood, and let household w be the last household for whom the benefits of broadband service are no less than $p(2)$. Then (3.b) is the second component of the algebraic description of situation 3.

$$s(2)[p(2)\text{-}c]w<K. \tag{3.b}$$

Situation 3 is the classic natural monopoly to which the classic price regulation policies, including rate-of-return regulation and price caps, might be applied. It should be noted, however, that the price reduction benefits associated with public ownership or public contributions toward common costs described for situation 2 might also be realized in situation 3. The difference is that such involvement by government is not needed to ensure that broadband service will be supplied. It should also be noted that price regulation can be a costly undertaking requiring resources and expertise unavailable to local governments. Price regulation may also affect investment incentives.

Situation 4: The neighborhood can support multiple suppliers, but not enough to drive price to the fully-competitive level.

Let n^* be the minimum number of suppliers required to drive the price of broadband service to its fully competitive level and let subscriber y be the last subscriber in the neighborhood valuing service at least as much as $p(n^*)$. Then conditions (4.a) and (4.b) describe situation 4.

$$s(2)[p(2)\text{-}c]w\geq K. \tag{4.a}$$

$$s(n^*)[p(n^*)\text{-}c]y<\mathrm{K}. \tag{4.b}$$

Because price falls with the number of competitors, the prospects for efficiency improvements from policy interventions are less promising for situation 4 than for situation 3, and are less so the greater is the number of competitors. The benefits of price caps in a multiform situation are conceptually the same as in a single firm situation, but competition among suppliers employing heterogeneous technologies might make the application of price caps to situation 4 neighborhoods problematic as underlying cost structures may differ among suppliers.

Situation 5: The neighborhood is supplied by enough firms to generate a fully competitive price.

Situation (5) is described by condition (5).

$$s(n^*)[p(n^*)\text{-}c]y>\mathrm{K}. \tag{5}$$

---

[8] $s(n)$ may be less than $1/n$ if early entrants retain persistent advantages in subscriber counts.

There is no obvious efficiency rationale for policies to promote penetration growth in type 5 neighborhoods.

Before concluding this subsection, we want to return briefly to the assumption that the cost of connecting a subscriber to the NDL is the same for all subscribers in a neighborhood that has been maintained from situation 2 on. High cost-high benefit subscribers may not be served under a single price regime if the broadband supplier has the option of refusing service to customers on which it loses money, even though net benefits for such potential subscribers may exceed the net benefits of service for subscribers with lower connection costs. While broadband providers currently are not precluded from charging higher prices to high cost customers than to low cost customers, the cost of negotiating one-off deals may preclude taking advantage of this option. On the other hand, if provision to such subscribers is mandated by policy, the excess of their connection cost over the common subscription fee must be made up through higher charges to other customers or through public support. These are long standing problems associated with universal service. Here we only want to point out that having to deal with variation in connection costs as well as variation in benefits received from service will add to the difficulties encountered in negotiating fees that vary among subscribers.

Network externalities can be built into this framework by adding a second benefit component, say β_i, reflecting the value of subscriber i to other broadband users, to the b_i that represents the benefits enjoyed by i. While it is often argued that such externalities are not incorporated in market prices, Panzar and Wildman (1995) show that this is not necessarily the case if each subscriber increases the use value of the network to other subscribers. The value of an externality is less likely to be incorporated in market choices if the benefits are more diffuse, such as the benefits to society as a whole of a better-informed citizenry.

B. The Economics of Aggregation at Higher Levels of the Network

In our stylized description of broadband networks, neighborhood feeder lines connect to community networks, which are facilities whose costs are common to all the neighborhoods wired for broadband in the community. Let F be the facilities cost of the community network and suppose there are h neighborhoods in the community for which the sum of net subscription benefits exceeds the cost of the NDL (i.e., condition (2.a) is satisfied.). Let B_j be the sum of net benefits for neighborhood j. Then the investment in the community network can be justified on efficiency grounds if condition (6) is satisfied.

$$\sum_{j=1}^{h} B_j - F \geq 0. \tag{6}$$

Conceptually, the issues raised by the need to cover common costs through charges levied on participants who may benefit differentially from using the network are the same at the community level as at the neighborhood level. Of course, neighborhood contributions to community level common costs must be collected through the fees charged individual subscribers. Ideally, these fees would be lower for neighborhoods for which the per subscriber difference between neighborhood net benefits and NDL costs are low than for neighborhoods for which this difference is large, but building such a relationship into a pricing formula may be difficult to accomplish. It is also possible that communities with neighborhoods that meet the positive net benefits minus NDL costs criterion for providing broadband service will not receive service because condition (6) is not satisfied. That is, once the costs of the community network are added in, the collection of neighborhoods making up the community may not collectively generate enough benefits in excess of their NDL costs to justify the expense of the community network. The likelihood of this outcome increases the greater the distance between neighborhoods and the smaller the number of neighborhoods making up the community.

It should be clear that the same issues arise when we move to the regional level of aggregation. Sparsely populated regions with substantial distance between communities may not generate sufficient subscriber benefits net of neighborhood and community common network costs to cover the costs of regional facilities. Even where the net benefit common cost comparison is favorable at the regional level, problems of allocation of regional common costs among communities and neighborhoods still remain.

Finally, we point out that the likelihood that the number of suppliers offering broadband service will vary among communities within a region and perhaps among neighborhoods within communities is a further challenge to the crafting of a coherent broadband policy. The fact that situation 2 and 3 challenges arise at the levels of communities and neighborhoods argues for a strong role for local governments and/or community organizations in developing initiatives tailored to local circumstances.

C. Promoting Higher Speed Networks

The arguments developed in the two preceding subsections assumed a single, fixed definition of broadband with performance characteristics dictated by market forces. Suppose it were federal policy to promote the development of a broadband infrastructure with higher bandwidth than the market would naturally offer. The fact that the higher bandwidth desired by policymakers did not arise naturally suggests that suppliers' forecasts of the incremental benefits of the added speed fall short of the incremental costs of providing it. (Or at least suppliers believe that the portion of the incremental benefits from higher bandwidth they can collect in higher subscription fees will be less than their costs for supplying it.) This suggests that an effect of mandating a higher bandwidth broadband service would be to convert some type 2 neighborhoods to type 1's, convert some type 3's to type 2's, and so on. In general, plans to promote higher speed networks than would be offered by markets on their own are likely to involve some sacrifice of competitive efficiencies and more direct government financial support to maintain or expand penetration.

V. Benefits and Broadband Network Requirements

The public policy debate is replete with claims as to the benefits of broadband networks and services although very few numerical estimates of expected benefits are available. Potential benefits of broadband are best measured relative to the status quo ante of electronic communications networks. One possible approximation of the benefits of broadband is the change in social surplus (consumer plus producer surplus) from the introduction of broadband technology platforms. Such benefits can originate from various sources. In so far as broadband networks allow the provision of existing services at lower costs, broadband will increase social surplus under given demand conditions. Compared with narrowband – dial-up or wireless – internet access broadband platforms also expand the set of feasible services and applications. In other words, they shift the production possibility frontier of an economy outwards. As new consumption opportunities are generated, social surplus is again enhanced.

Crandall and Jackson (2001) is the only paper that uses an estimate of consumer and producer surplus in assessing broadband. As some of the benefits of broadband will only materialize after the broadband market has reached a more mature growth phase, Crandall and Jackson base their analysis on a long-term perspective. The authors evaluate the benefits of broadband using two alternative approaches. The first approach uses a conjectured demand curve for broadband service assuming a monthly subscription charge of $40. The second approach examines specific applications and estimates the benefits accruing to consumers. Both approaches calculate benefits under two scenarios, a low estimate based on the assumption of 50% broadband penetration and a high estimate based on full diffusion of broadband, equivalent to the current penetration rate of voice telephone service. Based on the service specific approach, total annual benefits from all forms of broadband range from $272 billion in the partial diffusion to $520 billion in the universal diffusion scenario. Similar results, ranging from $297 billion to $460 billion annually are obtained from the conjectured demand function model.

Despite the advantages of a clearly defined concept of benefits, the vast literature on broadband typically does not use a stringent, operational definition. Nevertheless, strong arguments are still made in favor of policies facilitating a more rapid diffusion of broadband platforms. For example, in its report on electronic commerce the OECD (2000) concludes that the lead of the U.S. in this area is related to the wider availability of flat rate and always on internet access. Realizing the overall economic benefits from e-commerce would thus require policies facilitating always on access. Many more claims as to the beneficial effects of broadband are made in the context of specific applications. A survey of the pertinent policy documents and research literature by the authors of this chapter and two co-authors reveals that most of the potential benefits are expected in six categories of applications (Bauer, Gai, Kim, Muth and Wildman, 2002):

- Telemedicine
- E-learning
- E-government
- E-business
- Telecommuting
- Media and entertainment

However, a closer scrutiny of the claimed benefits reveals considerable variation in required bandwidth and that not all require high bandwidth platforms. Rather, in every category important choices exist and certain goals can be reached with different technological solutions. This point shall be illustrated for the first three areas.[9]

In the field of telemedicine, which is the provision of health care services, clinical information or medical education using telecommunications technology (Maheu et. al. 2001, Canadian National Broadband Task Force 2001), synchronous and asynchronous services need to be distinguished. Bandwidth requirements vary widely with the service provided and not all applications require broadband. Clinical applications include telediagnosis, telemonitoring, teleconsultation, teleradiology, the use of remote medical instruments, and behavioral healthcare mediated through telecommunications. In addition to clinical applications, information and communications technology can reduce the administrative costs of health care providers and facilitate continuing education programs. Health care professionals distinguish between different levels of telemedicine requiring different network support. Table 4 illustrates the different telecommunications needs of selected applications. It is clear that only the more advanced applications require broadband platforms. This threshold may shift upward with more sophisticated compression techniques.

Table 4. Forms of telemedicine and their bandwidth requirements

Applications	Minimum Network Requirement	Supplementary Devices / Resources
Telediagnosis	POTS	-Telephone, mail, email or fax
Teledermatology	POTS	-Telephone, mail, email or fax, video display units with high resolution image data if necessary
Teleultrasound (Telesonography)	-4 or 6 ISDN lines for cardiology real-time motion video (256 or 384Kbps)	1. Experts with extensive information regarding the overall state of the patient, moderate operator, and an expert for reading images at once 2. Devices, PACS, MM tools, video display units, audio & video codecs, VC conference management tools, electronic board, microphones, speakers, ultrasound device, remote control, control board
Telemonitoring	-56K, BRI ISDN, fractional T-1 -ATM, DSL in case real-time multimedia communication is needed	1.An expert who monitors the sent biosignals and image data 2. Devices: blood pressure monitors, pulmonary function monitors, electronic stethoscope, tele-EKG (Electrocardiogram), tele-echocardiography and ultrasound systems, untegrated light, video camera platforms, dermascope, 1-CCD camera, dental cameras
Teleconsultation	-At least fractional T-1 -ATM, DSL or cable network	1.High resolution visual data and video display units for patient monitoring 2. Videoconferencing systems and interactive video room systems 3. Computer-supported teleconferencing systems
Teleradiology	-ATM	-PACS (Picture Archiving and Communication Systems), high-resolution monitors, connectivity system, frame grabbers, compression techniques for heavy loaded image or video data, store-and-forward systems (audio and video capture card, camera, microphone, and image management software)

E-learning also can be configured in various ways. Asynchronous e-learning uses web-based learning modules but does not support real time interaction between the instructor and the students. Synchronous e-learning consists of on-line real-time lectures, which typically have to be joined by students at the time of their delivery. Additional asynchronous functions typically support the learning environment. Most demanding in terms of bandwidth are forms

[9] For a more detailed discussion of e-business, telecommuting, and media and entertainment. see Bauer, Gai, Kim, Muth and Wildman 2002.

of collaborative e-learning in which students have to interact continuously to solve problems or engage in other learning activities. Table 5 illustrates that only certain forms of e-learning require broadband support.

Table 5. Broadband requirements of e-learning

	Applications	**Network Demands**	**Complementary Functions & Tools**
Asynchronous	Computer-Based Training Modules (CBT), Multimedia Database Support System (MDS)	POTS, ISDN	Email, automatic upload of educational materials
Synchronous	Remote Lecture Room (RLR), Interactive Home Learning (IHL)	Up to 6 ISDN channels, ATM, Internet protocol stack	Bulletin board, videoconference or audio conference systems , email, chat room, file exchange tool
Collaborative	Remote Interactive Seminars (RIS)	Up to 6 ISDN channels, ATM, Internet protocol stack	Bulletin board, videoconference or audio conference systems , email, chat room, file exchange tool

E-government is another application that is often portrayed as dependent on the availability of a broadband infrastructure. E-government includes the use of automated systems in transactions between the government and the public as well as the exchange of information between governments. Local, state and the federal governments have started to use the Internet to better communicate with citizens. Applications include the use of websites to make information available to the public, the utilization of the internet to make forms and applications available, and the use of the internet for transactions (e.g., drivers license applications, payment of tickets, or the filing of taxes). In addition, some agencies have begun to use videoconferencing to conduct proceedings with remote stakeholders and there are discussions to establish tele-courts. Table 6 summarizes the technological requirements for different kinds of applications. As in the other areas it is clear that not all e-government applications require broadband.

Table 6. E-government applications and bandwidth requirements

e-Government Applications	Technology	Network Implications	Bandwidth requirement
General information publication, search and retrieve	Web browsing, file downloading and uploading (HTML, HTTP, FTP, Telnet, database, etc.)	Fast web page downloading and file transferring	-Low bandwidth requirement for individual users -High bandwidth and capacity requirement for government networks
G2C transactions (ticket payment, bill payment, tax payment, permit application, filling of licenses, online crime reporting, online Citizens complaint, Street light repair request, online voting	Web browsing, Web interactions (HTML, HTTP, FTP, Telnet, Java, database etc.)	-Fast web page downloading and file transferring -Instant response required by real-time interactivity -Security and privacy	-Low bandwidth requirement for individual users -High bandwidth and capacity requirement for government networks
Multimedia Applications (Webcasting, 3D navigation, high definition pictures and video, IP telephony)	Streaming media, VRML, HTTP, etc.	-Fast data transmission to ensure video and audio quality, interactivity -Low latency and jitter	High bandwidth connections are required or strongly recommended
G2B transactions (Government e-Procurement, G&B collaboration)	EDI (Electronic Data Interchange), Integrated voice, data, and video transmission	-7 days/week, 24 hours/day availability -Time-critical interaction and response -Security	High bandwidth requirement; broadband is an enabling factor
G2G (inter-government information sharing and transactions, video conferencing) and Internal efficiency and efficacy (Intra-government information publication and sharing, internal daily operations)	Integrated voice, data, and video transmission	-7 days/week, 24 hours/day availability -Time-critical response -Security -Flexibility	High bandwidth requirement; broadband is an enabling factor

In addition to demonstrating the very disparate bandwidth requirements of different services identified with broadband, the telemedicine and e-learning examples illustrate that, while broadband technology is general purpose, some applications seen as important may useful to fairly narrow subsets of broadband users. Furthermore, the individuals jointly utilizing these services may be located quite far from each other. This analysis provides an important basis for the discussion of public policy issues. It is a first indication that it is probably necessary to adopt a differentiated view of broadband networks. The best networking solution is very likely dependent on local conditions, the overall vision of a community, and the locations and sources of opportunities for collaboration on specific applications with other users and institutions.

VI. Policy Responses to the Spatial Character of Benefits and Costs

The need for individuals and institutions in different locations to coordinate activities and to collaborate on investments in complementary assets and human resources to realize the benefits of certain broadband applications means that policies promoting these applications must explicitly consider the spatial character of benefits and costs. In general, one would expect coordination costs to increase with the distance over which coordination is required. Commitments to services requiring coordination among users and especially across substantial distances can easily be stalled by the chicken-and-egg problem of who moves first. The need for coordination among users also means that the benefits calculations of individual adopters underlying the analysis of Section 4 may be influenced by coordination failures and successes.

From an analytical perspective it is necessary to explore whether decentralized decision-making by individual firms and potential users of broadband is capable of producing a socially optimal outcomes. In the light of our discussion of the differentiated network requirements of envisioned broadband services it seems helpful to further examine the nature of the benefits from broadband service as well as their spatial characteristics. Such an analysis reveals that the potential benefits from broadband service indeed have very different characteristics. Some benefits, such as more efficient procurement or reduced costs for certain health care services, are essentially private goods that can be fully internalized by the users. Other benefits, such as high speed access and support services in a business park may have the character of club goods. That is, they benefit a group of stakeholders and will only be supplied if that group acts collectively. If broadband benefits are such club goods, they will only be forthcoming if the "club" is formed. However, if the required "club" is established, all benefits are internal to its members. Finally, benefits may be pure public goods, accruing to a community or several communities at large. For example, the availability of advanced distance education technology may reduce the crime rate in a community and enhance the quality of life even for those who do not use distance education. In this latter case, the immediate users of broadband services cannot easily internalize the benefits and public policy action may be necessary.

Table 7. Spatial nature of broadband costs and benefits

		Benefits		
		Local	**Inter-local**	**Global**
Costs	**Local**	E-government	Tele-education	Tele-education
	Inter-local		Tele-medicine	Tele-medicine
	Global			E-business

In order to design appropriate policy intervention a better understanding of the spatial structure of costs and benefits from broadband networks is also necessary. Costs and benefits can be local in scope, they can be inter-regional or they can be global. This yields a total of nine possible scenarios, which are synthesized in table 7. If both the costs and benefits of broadband are local, policy decisions are best taken at the local level. If benefits and costs are inter-local or global, unique problems arise, as optimal policy will require coordination among the stakeholders in different communities. As long as the transaction costs of achieving such coordination are lower than the net benefits of broadband to all stakeholders, such an approach will likely be forthcoming. However, if coordination is complex and its transaction cost above the net benefits of broadband, voluntary coordination may not result in broadband investment. In such cases, coordination at a higher level of the political system may provide a solution. Additional problems arise in asymmetric cases, for example, when the costs of broadband are local but the benefits inter-regional or global. In the absence of transaction costs, optimal outcomes could be achieved regardless. However, under real world conditions the level of broadband investment will be shaped by the ability of the stakeholders to internalize effects through negotiation and coordination. Again, policy-making at a higher level in the political hierarchy may offer a feasible solution.

VII. Summary and Conclusions

The shared cost characteristics of broadband networks suggest that broadband policies will be most effective if they are tailored to local market conditions. However, the fact that sharing occurs at successively higher levels of the network suggests that region-wide coordination for policy development may be best for large areas with low population densities. Our analysis also shows that policies that promote higher bandwidth services than markets offer on their own likely will reduce the number of local markets with effective competition and will expand the role of government in promoting broadband deployment in many areas.

Broadband policies should be tailored to the spatial and product characteristics of broadband services. In the case of predominantly private benefits or club goods, the role of the public sector will best be focused on measures facilitating demand synchronization and aggregation. Where the mix of benefits has the character of public goods, other means might be more effective, including public funding or even direct provision. The majority of present broadband policy measures are not based on a systematic consideration of these issues. However, many of the practical policy measures explicitly or tacitly recognize at least some of the spatial aspects of the problem. For example, the LinkMichigan program encourages communities to plan broadband networks jointly. Such an approach may be justified if the costs and benefits of broadband reach beyond local communities.

REFERENCES

APT (2002), Advanced services, enhanced lives: an examination of broadband application case studies and policy recommendations to accelerate deployment. Presented to the Broadband Forum, February 8, 2002, Washington, DC: Alliance for Public Technology.

Aron, D. J. and Wildman, S. S. (1996), "The Pricing of Customer Access in Telecommunications," *Industrial and Corporate Change* 5(4), 1029-1047.

Bauer, J. M., Gai, P., Kim, J., Muth, T. A. and Wildman, S. S. (2002), *Broadband: benefits and policy challenges*, research report, Quello Center at Michigan State University, August 2002.

Canadian National Broadband Task Force. (2001), *The new national dream: networking the nation for broadband access*, report of the National Broadband Task Force, Ottawa.

Crandall, R. W., and Jackson, C. L. (2001), *The $500 Billion Opportunity: The Potential Economic Benefit of Widespread Diffusion of Broadband Internet Access*. Criterion Economics, L.L.C.

Faulhaber, G. R. and Hogendorn, C. (2000), "The market structure of broadband telecommunications," *Journal of Industrial Economics* 48(3), 305-29.

Federal Communications Commission, (2002). "High Speed Services for Internet Access: Status as of June 30, 2002."

Firth, L. and Longstaff, P. (2002), "Convergence," paper presented at the 14th Biennial Conference of the International Telecommunications Society, Seoul, Korea, August 18-21, 2002.

ITU (February 25, 2003 update), "Top 15 Economies by Broadband Penetration," International Telecommunications Union: http://www.itu.int/ITU-D/ict/statistics/at_glance/top15_broad.html.

Maheu, M. M., Whitten, P., and Allen, A. (2001), *E-health, telehealth, and telemedicine: A guide to start-up and success*. San Francisco: Jossey-Bass.

OECD (2000), *Local access pricing and e-commerce*, Paris: Organisation for Economic Co-operation and Development.

OECD. (2002). *Broadband infrastructure deployment: The role of government assistance*, Paris: Organisation of Economic Co-operation and Development.

Panzar, John C. and Wildman, Steven S., "Network Competition and the provision of universal service," *Industrial and Corporate Change* 4(4), 711-719.

Pepper, R. (2002), "Competition, regulation and digital migration," paper presented at the 14th Biennial Conference of the International Telecommunications Society, Seoul, Korea, August 18-21, 2002.

INDEX

H

I

K

L

M

N